Kazakhstan's Developmental Journey

Anastasia Koulouri · Nikolai Mouraviev
Editors

Kazakhstan's Developmental Journey

Entrenched Paradigms, Achievements, and the Challenge of Global Competitiveness

palgrave
macmillan

Editors
Anastasia Koulouri
University of East Anglia
Norwich, UK

Nikolai Mouraviev
Abertay University
Dundee, UK

ISBN 978-981-15-6898-5 ISBN 978-981-15-6899-2 (eBook)
https://doi.org/10.1007/978-981-15-6899-2

This Palgrave Macmillan imprint is published by the registered company Springer Nature Singapore Pte Ltd.
The registered company address is: 152 Beach Road, #21-01/04 Gateway East, Singapore 189721, Singapore

Anastasia Koulouri
To my mother and sister,
I couldn't do it without you!
With all my love, thank you!

Contents

Notes on Contributors

Slyamzhar Akhmetzharov is a Ph.D. candidate at the Graduate School of Public Policy, Nazarbayev University, Nur-Sultan, Kazakhstan. He has worked as a consultant for governmental and private think tanks. His research interests include informal relations in the public sector, legal reforms, and protest actions. Currently, he works on a project aiming to identify the factors leading to labor conflicts in extractive industries in Kazakhstan.

Aziza Baubekova is a Teaching Fellow in Biology at the Center for Preparatory Studies, Nazarbayev University, Nur-Sultan, Kazakhstan. She has recently joined the Water, Energy, and Environmental Engineering Research Unit at the University of Oulu, Finland, as a Ph.D. candidate. Her work is interdisciplinary and focuses on the water-energy-food nexus in Central Asia including water security, climate change, and environmental modeling.

Dr. Aigul Beimisheva is an Associate Professor at the Department of Sociology and Social Work, Narxoz University, Almaty, Kazakhstan. She has 7 years of international and national experience in the planning and management of social and sociological research. She holds a Ph.D. in Sociology from the Al-Farabi Kazakh National University, Almaty, Kazakhstan.

Aliya Bizhanova is a Research Grant Coordinator and a Senior Lecturer at the Department of Research and Development, Almaty Management

University (AlmaU), Almaty, Kazakhstan. She is also an OECD consultant for the National Skills Strategy. Between 2012 and 2019 she was an operations officer for the World Bank Education Global Practice. She holds a master's degree in International Education Policy from Vanderbilt University (U.S. Department of State, E. Muskie fellowship), and a diploma of higher education in Law from Samara State University, Samara, Russia.

Shaimerden Chikanayev is an advocate of the Almaty City Bar and a partner of GRATA International Law Firm, Almaty, Kazakhstan. He has over 13 years of experience in Kazakhstan's legal services market, and significant experience in matters relating to countries of Central Asia. His practice focuses on a wide range of finance and mergers and acquisitions transactions. Shaimerden has extensive knowledge of various industries, and in particular energy, utilities and social infrastructure, financial services, and transportation.

Dr. Saule Emrich-Bakenova is an Associate Professor at the Department of Public Administration and International Development, KIMEP University, Almaty, Kazakhstan. She has published in international refereed journals in the areas of public policy and public administration. She also has several years of policy consultancy experience.

Dr. Madina Junussova is a Research Fellow at the Institute of Public Policy and Administration of the University of Central Asia, Kazakhstan/Kyrgyzstan/Tajikistan, and a CERGE-EI Foundation Teaching Fellow. She has 15 years of international and national experience in the planning and management of urban and regional development. She is a member of the International Public Policy Association (IPPA), and the International Society of City and Regional Planners (Isocarp). She holds a Ph.D. in Public Policy from Carleton University, Ontario, Canada.

Dr. Marat Karatayev is an Ernst Mach Researcher at the Institute for Systems Science, Innovation and Sustainability Research, Karl-Franzens University of Graz, Austria, and a Research Fellow at the Institute for Advanced Studies on Science, Technology, and Society, Graz University of Technology, Graz, Austria. He holds a master's by research degree in Environmental Protection Engineering, and a Ph.D. in Urban, Energy and Environmental Planning. His research has focused on the political economy of energy transitions, innovations and climate change, water-energy-food nexus governance, and scenario generation tools.

Dr. Anastasia Koulouri is a Lecturer at the Norwich Business School, University of East Anglia, Norwich, UK. Previously she has held posts at the University of Strathclyde and Abertay University (UK), the Technical University of Helsinki (Finland), and the Belgian Nuclear Energy Research Centre (Belgium). She holds a Ph.D. in Management Science from the University of Strathclyde, Glasgow, UK. Her research interests include energy policy and security in resource-rich countries with particular focus on Kazakhstan, and sustainable development of transitional economies. Dr. Koulouri has co-edited volumes on *Energy Security* (Palgrave Macmillan, 2019) and on the *Water-Energy-Food Nexus* (Routledge, 2019).

Anastasia Kvasha is a Teaching and Research Assistant at the Environmental Systems Laboratory, Central European University (CEU), Budapest, Hungary. Her work is focusing on applications of space-based earth observations for disaster risk reduction and climate change adaptation. She has been a consultant at the Food and Agriculture Organisation of the United Nations (FAO) and a civil servant in Russia. She is currently pursuing a Ph.D. at the CEU.

Dr. Nikolai Mouraviev is a Senior Lecturer at the School of Business, Law and Social Sciences, Abertay University, Dundee, Scotland, UK. Previously he has held teaching positions at KIMEP University (Kazakhstan); Wayne State University (US); and Viterbo University (US). Part of his research focuses on public-private collaboration in developing nations, and he has published extensively in this area including a monograph (Palgrave Macmillan, 2017) and an edited volume (Cambridge Scholars Publishing, 2017). In addition, his research interests include entrepreneurship and governance of the energy sector. Dr. Mouraviev has co-authored seven books, including two research monographs, and published more than 50 papers.

Dr. Serik Orazgaliyev is an Assistant Professor at the Graduate School of Public Policy, Nazarbayev University, Nur-Sultan, Kazakhstan. He has been a visiting academic at the Lee Kuan Yew School of Public Policy, National University of Singapore, Singapore, and a visiting research affiliate at the Cambridge Central Asia Forum, University of Cambridge, UK. He is the recipient of the Bolashak International Scholarship and the Asian Universities Alliance Scholars Award. Dr. Orazgaliyev's research interests include governments and multinational enterprises, institutions and development policies, and international political economy.

Dennis Soltys is a Professor at the Department of Public Administration and International Development, KIMEP University, Almaty, Kazakhstan. Professor Soltys has a master's degree in Soviet and East European Studies from Carleton University, Ontario, Canada, and a Ph.D. in Political Sciences from the University of Toronto, Canada. His main research interest is the historical and cultural foundations of comparative educational systems.

Dr. Anastasia Tikhonova is a Teaching and Research Fellow at the University of Yangon, Myanmar, where she teaches climate change, and water and food security courses. Previously she has worked in the Food and Agriculture Organisation of the United Nations (FAO) and the Leibniz Institute for Agricultural Development in Transition Economies (IAMO). She has a Ph.D. in Environmental Sciences and Policy from the Central European University, Budapest, Hungary, and a vast research experience in the region of Central Asia.

List of Figures

List of Tables

CHAPTER 1

Introduction: An Overview of Kazakhstan's Developmental Journey Since 1991

Anastasia Koulouri

Kazakhstan: An Overview

Kazakhstan is a landlocked, transcontinental country with the majority of its territory in Central Asia and only the most western parts located in Eastern Europe. It borders with Russia, China, Kyrgyzstan, Uzbekistan, and Turkmenistan. It is the ninth largest country in the world in terms of land area, covering 2.725 million km^2 (Worldatlas 2020), and, with an estimated population of 18.73 million (Worldometers 2020), has one of the lowest population densities in the world (6–7 people per km^2).

Brief History of the Nation

The country has a diverse cultural heritage encompassing 130 ethnicities, including, in 2017, 67% of ethnic Kazakhs, 20.2% of Russians, and 3.1% of Uzbeks (Encyclopædia Britannica 2020). Modern-day Kazakhstan, formally the Republic of Kazakhstan, emerged from the settlement of

A. Koulouri (✉)
University of East Anglia, Norwich, UK
e-mail: a.koulouri@uea.ac.uk

A. Koulouri and N. Mouraviev (eds.),
Kazakhstan's Developmental Journey,
https://doi.org/10.1007/978-981-15-6899-2_1

nomadic tribes, several waves of immigration, and a number of administrative reorganisations. In the mid-nineteenth century, the area currently occupied by Kazakhstan was nominally ruled by the Tzar as part of the Russian Empire. Following the Russian Revolution of 1917 and the ensuing civil war, Kazakhstan went through a turbulent transition including a brief period of independence. In 1936, the Kazakh Soviet Socialist Republic was created as part of the Soviet Union, and, in 1991, Kazakhstan declared its independence. Since then, and until his resignation in March 2019, the nation has been headed by the First President, Nursultan Nazarbayev, who now is the Chairman-for-Life of the nation's Security Council. President Nazarbayev was one of Kazakhstan's leaders during the Soviet era and has been guiding the nation through the challenging, post-independence period. As per the constitution, President Nazarbayev was succeeded by the Senate Chairman, Kassym-Jomart Tokayev, who was elected President in June 2019.

Public Administration

Kazakhstan is a presidential constitutional republic and a unitary, secular state. The government consists of the Prime Minister, three Deputy Prime Ministers, 16 Ministers, and officials including those of the Offices of the President and the Prime Minister. The government is formed by the President, who is also the Commander-in-Chief of the armed forces, and can veto legislation passed by the parliament. Kazakhstan has a bicameral parliament consisting of the lower house, the *Majilis*, and the upper house, the Senate. Administratively, the country is divided into 14 principal administrative regions plus two "cities of republican importance" (each of these cities has the status of a region), the capital Nur-Sultan (previously Astana) and, the biggest city and old capital, Almaty. Each of these regions (*oblast*) is led by an *Akim*, appointed by the President. The regions are subdivided into 117 districts (*Rayons*), each led by a district *Akim*, appointed by the corresponding regional *Akim*.

Economic Development

Consequently, the government, including budget allocation, is centralised. There is therefore, to a certain degree, a continuation of the government model in place during the Soviet era. At the same time, since 1991, the country has been transitioning from a planned to a market economy. The extent to which this transition is complete is beyond the scope of this book. Nonetheless, it is worth noting that, in March 2002, the US Department of Commerce awarded Kazakhstan the status of a "market economy" (Eicher 2004), and, in November 2015, Kazakhstan acceded to the World Trade Organisation (Export.gov 2019). More recently, in 2019, Kazakhstan was ranked 55th (among 141 countries) on the Global Competitive Index 4.0, which "measures national competitiveness – defined as the set of institutions, policies and factors that determine the level of productivity" in a nation (Schwab 2019, p. xiii). However, the World Bank ranked Kazakhstan 25th (among 190 economies) in terms of the ease of doing business in the country (World Bank 2020a), with a score equal to that of Canada and Iceland.

Kazakhstan has an abundance of natural resources, including oil, gas, coal, uranium, and other minerals. Largely dependent on its extractive industries, and in particular oil and gas, Kazakhstan has become the largest and strongest economy in Central Asia. Its GDP in the first quarter of 2020 was USD 170.9 billion, and, in 2019, its growth rate was 4.5% and the rise in investment 12.9% (World Bank 2020b). Moreover, Kazakhstan has emerged as a transport and logistics hub, linking the markets of China and Southeast Asia to those of Russia and Europe (World Bank 2020b; Export.gov 2019). The nation's exports of oil, gas, and their products represented 21% of its GDP in 2018 (World Bank 2019) and 67% of its exports in 2019 (Trading Economics 2020). This dependence has left the country vulnerable to oil price fluctuations and market volatility, while global recessions have also negatively affected the nation's economy.

Strategic Direction

To support its path to established statehood and economic growth, Kazakhstan has set its priorities in the strategies *Kazakhstan-2030* and *Kazakhstan-2050*, introduced in 1997 and 2012, respectively (Nazarbayev 1997, 2012). These strategies outline the nation's development plans until 2030 and 2050, respectively. To operationalise these strategies, numerous legislative acts, changes to existing legislation, and state programmes have been introduced.

Kazakhstan-2030 focuses on "transforming the country into one of the safest, most stable, ecologically sustained states of the world with a dynamically developing economy" (Nazarbayev 1997). Its primary aim was to restructure the economy and exploit the potential of non-primary sectors, such as agriculture and manufacturing, by attracting foreign investment, while developing the necessary human capital, improving the nation's infrastructure, modernising and streamlining the government, reforming public service provision, and placing emphasis on the health and well-being of its citizens (Nazarbayev 1997).

In 2012, First President Nazarbayev declared that Kazakhstan had reached the goals set in *Kazakhstan-2030* ahead of schedule and had been established as a modern state. Recognising ten global challenges of the twenty-first century,[1] he introduced the *Kazakhstan-2050* strategy, setting the nation's future path with the overarching aim that Kazakhstan must enter the top 30 most developed countries in the world by 2050. The strategy's seven key objectives (Nazarbayev 2012) are:

1. transitioning to a new economic policy path based on competitiveness, profitability, and return on investment by improving the investment environment, promoting industrialisation with emphasis on export-oriented non-energy sectors, and fostering technological innovation while ensuring that the nation's future energy, water, and food needs are met;
2. providing extensive support for entrepreneurship as the driving force of the economy;
3. modernising the education system with particular focus on ensuring that current and future demands of the economy are met;
4. developing a new social model by introducing new principles of social policy, and modernising the labour policy and the healthcare system;
5. strengthening the nation's statehood and consolidating the development of democracy by decentralising power, creating a professional state apparatus, strengthening the links between the state and business, and fighting corruption;
6. promoting regional and national security and stability while pursuing the nation's interests by setting a "consistent and

[1] These are the accelerating course of history, the global demographic imbalance, food security, water shortages, global energy security, the depletion of natural resources, the third industrial revolution, growing social instability, the crisis of our civilisation's values, and the continued global destabilisation (Nazarbayev 2012).

predictable foreign policy", actively developing economic and trade diplomacy, and intensifying international cooperation;
7. safeguarding the nation's "multi-ethnic and multi-confessional society" by strengthening the nation's identity.

The Collection

The editors of this collection have many years of work experience in Kazakhstan on which the book draws. Initially focusing on energy policy and, in particular, on renewable energy (Koulouri and Mouraviev 2018; Mouraviev and Koulouri 2017, 2018, 2019b), the thematic focus of the editors' work later progressed to encompass energy security: its understanding by academics and governments, and its conceptualisation by the latter as manifested by the nation's energy policy (Koulouri and Mouraviev 2019; Mouraviev and Koulouri 2019a, c). Most recently, this work expanded to consider the role of a water–energy–food (WEF) nexus conceptualisation in the policy and governance of the WEF sectors within the context of sustainable development (Koulouri and Mouraviev 2020a, b). To do so, the lens of relational equity management was utilised, providing an integrative approach that permits the holistic consideration of interlinkages, interdependencies, synergies, and trade-offs in policy and governance of the WEF sectors.

Focusing on energy policy, energy security, and a nexus conceptualisation for policy and governance in the WEF sectors has been useful. However, it provided a relatively narrow perspective as it did not permit to draw insights into the evolutionary journey of Kazakhstan since it declared its independence in 1991, and its trajectory from a planned to a market economy. With the introduction of the *Kazakhstan-2030* and the *Kazakhstan-2050* strategies, a number of questions have arisen as to the suitability of policy initiatives and governance actions in key areas and sectors, as well as related paradigms, to pursuing the nation's strategic objectives. Do the current policies, governance mechanisms, and paradigms fit the *Kazakhstan-2050* strategy? Are they effective for the nation's long-term economic and social development? What is their contribution to the nation's sustainable growth? What should an ideal contribution be? Should there be a policy and/or a paradigm shift, and how should this materialise? What conditions should be formed to ensure this shift, and what are the potential impediments in creating them? What are the key actors with a significant impact on the desired shift, and how can they be motivated to collaborate? What are the potential barriers in their collaboration, and how could these be mitigated, overcome, and/or removed? Answering these questions has been the motivation in compiling this edited collection and has informed its thematic focus.

The book's *aim* is to discuss the transitioning trajectory of Kazakhstan since it declared its independence from the Soviet Union in 1991 to the present day by analysing the evolution of key policy areas and sectors through the lens of policy development and implementation. The book charts and assesses the policy and governance development, and evaluates their appropriateness for the pursuit of the country's strategic objectives. By assessing policies against current conditions in the selected policy areas and sectors, the book highlights achievements, existing and missed opportunities, and policy design and implementation challenges to Kazakhstan's development. Furthermore, the book draws together insights into the suitability of entrenched, evolving, and nascent policy paradigms to the political, economic, and social imperatives of the twenty-first century.

The book's *thematic focus* is Kazakhstan's evolutive journey viewed through the critical examination of the elaboration of policies, governance schemes, and policy paradigms, in terms of their fit with the *Kazakhstan-2050* strategy and their effectiveness in ensuring the nation's long-term, sustainable, socio-economic development and growth. Should policies, governance mechanisms, and/or paradigms be found to be inadequate, a shift in policy and/or the relevant paradigms, and/or changes to governance might be necessary to provide the required impetus. Successfully enacting such a shift would require the creation of the necessary conditions, and the involvement of and contribution by key actors, motivated to collaborate and constructively engage in the process.

The book's *primary themes* examine areas that have received insufficient attention to date, and explore critical questions relating to Kazakhstan's transitional journey, including:

- the role of key policy initiatives to develop and restructure the nation's economy, and the extent to which these attempts have been successful in fulfilling their goals, or have acted as barriers in achieving long-term economic sustainability (Chapters 2 and 3);
- the emergence of new policy paradigms in delivering public services and improving the nation's infrastructure while addressing the challenges of corruption and fairness, inefficiency and innovation, and public funding limitations (Chapters 5 and 6);

- the function of educational communities as a catalyst for delivering the nation's higher education policy within the context of developing its human capital and, thus, advancing Kazakhstan's transformation into one of the 30 most developed nations in the world (Chapter 4);
- the necessity of adopting WEF nexus thinking in policy design, development, and implementation for the WEF sectors, and, thus, fulfilling in a sustainable manner the nation's water, energy, and food needs (Chapter 7).

This is a non-technical book that considers the selected policy areas and sectors through the lens of policy development and implementation seeking to critically review their effectiveness in achieving Kazakhstan's stated strategic objectives. *Kazakhstan-2050* does recognise the importance of technological innovation, research and development, and the need to identify and exploit technological trends with high potential. Nonetheless, the successful implementation of policies has to be underpinned by organised actors, funding, institutions, procedures, mechanisms, and tools, while a focus on technological innovation does not necessarily lead to appropriate policy and governance solutions, and is insufficient in itself to successfully pursue policy objectives.

The *main argument* of the book is that Kazakhstan's policies for key areas and sectors, developed in the 1990s and 2000s, have to date served the nation's needs. Nonetheless, the country is seeking to achieve a competitive edge on the world stage (Nazarbayev 2012). In this context, many of the nation's policies indicate a need for further elaboration and progression, or a significant shift of their underlying paradigms. New policy paradigms, coupled with innovative governance tools, may accelerate the nation's modernisation and bring long-desired and highly anticipated reforms, enhancing the new political leadership's credibility, and ensuring the nation's long-term sustainable growth.

The Book's Structure

The book includes three parts, offering a blend of chapters that discuss the role of key sectors in Kazakhstan's developmental journey from 1991 to date, and chapters that look beyond the boundaries of specific sectors, exploring emerging paradigms in key policy areas within the context of a growing market economy.

The first part of the book includes Chapters 2–4. They discuss the role of the oil sector in the nation's transitional journey and the extent to which its natural resource abundance enables or hinders its sustainable economic growth; the evolution of policy and governance in the agricultural sector within the context of its significance for the country as a key non-energy sector providing employment for a large part of the population, and securing the nation's future food needs; and the function of educational communities in realising the nation's higher education policy through increased autonomy, the ability to self-organise, and enhanced capacity to influence policy and governance.

The second part of the book (Chapters 5–8) offers a critical examination of the emergence of new, and the transformation of established, paradigms. It includes chapters that discuss the elaboration of the public administration paradigm underpinning the delivery of public services within the context of transparency and procedural fairness; the emergence of public–private partnerships as a new paradigm for the provision of public services, the development of infrastructure, and the strengthening of the links between the state and the private sector; the extent to which it is imperative to embed a WEF conceptualisation in policy and decision-making regarding the water and energy sectors, and in agricultural development; and the evolution of policy on small towns that depend on a single local industry as a means to drive social reform and address regional inequalities in Kazakhstan.

It is unfeasible for a single book to consider all policy areas and sectors, and this is indeed beyond the scope of this book. Rather, this edited collection, by its selection of policy areas and sectors, endeavours to provide illustrations of Kazakhstan's developmental journey that offer the reader a significant breadth of perspective, and an insightful glimpse into the nation's transition to a market economy.

The third part of book (Chapter 9) explains how ideas offered by the book's chapters link to each other and, importantly, what one can learn from them. The conclusion identifies the implications for Kazakhstan's future policy and governance in the light of the ambitious *Kazakhstan-2050* strategy, and draws insights into themes that have been reiterated by the authors throughout the book, such as sustainability and economic restructuring.

The book's *original and distinctive contribution* to the literature on Kazakhstan is the insights it provides into its transformative trajectory drawn by critically examining the appropriateness of policies and relevant paradigms to the evolutive needs of key policy areas and sectors within the context of the country's growth priorities. The book provides a novel outlook on Kazakhstan's transition through the lens of the evolution of policy, governance, and policy paradigms, which differentiates it from competing titles. Moreover, the book calls attention to successes and enduring challenges, and elucidates current and missed opportunities in policy and governance. Finally, the book illuminates entrenched and emergent policy paradigms in Kazakhstan, and assesses their adequacy in view of the socio-economic and political challenges of the twenty-first century.

The Book's Readership

By its unique and distinct contribution to the body of knowledge on Kazakhstan, the book will satisfy the needs of a broad readership—academics, undergraduate and postgraduate students, practitioners, and all those with an interest in Central Asia and/or Kazakhstan.

In particular, the book will benefit researchers, teaching academics, and students interested in policies in the post-Soviet era, and seeking a critical review of policy and governance in key areas and sectors in Kazakhstan from 1991 to date. Moreover, it will be useful to academics and students (primarily postgraduate) studying public policy and governance, public administration, transition to a market economy, sustainable development, and regional studies focusing on Central Asia and/or Kazakhstan. By reading this book, academics and students may identify enablers and barriers in the development of policy design and implementation in the specific geopolitical and socio-economic context of post-Soviet Kazakhstan. Building on this, readers could compare and contrast the evolutionary paths of Kazakhstan with those of other Central Asian ex-Soviet republics, and draw insights into their experience and achievements, and the challenges they faced, and continue to face, in transitioning to a market economy, and how these were overcome.

Further, the book may be of interest to practitioners, such as policy designers and regulators, government experts, and consultants, working in Central Asia and/or Kazakhstan. It may also be of interest to managers and project staff from international and national agencies, and non-governmental organisations involved in sustainable development and resource management in the region.

These practitioners may find the edited collection of interest as it highlights a multitude of challenges and opportunities, barriers and enablers that the government of Kazakhstan may face while transitioning from a centrally controlled to a market economy. This analysis and critical appraisal through the lens of policy and governance will be of interest to practitioners worldwide enabling them to compare and contrast the account provided by the book with their own appreciation of the critical elements of policies and governance tools for the successful transition to a market economy, ensuring the long-term, sustainable development of a nation. Further, practitioners may identify commonalities and differences in how policy design, development, and implementation evolved during Kazakhstan's journey when compared with their experience from other ex-Soviet, Central Asian nations, allowing them to identify best practice and learn from Kazakhstan's experience.

References

Eicher, S. (2004, April). *Is Kazakhstan a market economy yet? Getting warmer...* (William Davidson Institute Working Paper 673). Available via https://papers.ssrn.com/sol3/papers.cfm?abstract_id=539702. Accessed 2 May 2020.

Encyclopædia Britannica. (2020). *Kazakhstan: Ethnic composition, 2017*. Encyclopædia Britannica. Available via https://www.britannica.com/place/Kazakhstan/People#/media/1/313790/236313. Accessed 24 April 2020.

Export.gov. (2019, June 28). *Kazakhstan-Market Overview 2019*. Export.gov. Available via https://www.export.gov/article?series=a0pt0000000PAu9AAG&type=Country_Commercial__kav. Accessed 23 April 2020.

Koulouri, A., & Mouraviev, N. (2018). Governance of the clean energy sector in Kazakhstan: Impediments to investment. *International Journal of Technology Intelligence and Planning, 12*(1), 6–23.

Koulouri, A., & Mouraviev, N. (2019). Energy security through the lens of renewable energy sources and resource efficiency. In N. Mouraviev & A. Koulouri (Eds.), *Energy security: Policy challenges and solutions for resource efficiency*. London: Palgrave Macmillan.

Koulouri, A., & Mouraviev, N. (Eds.). (2020a). *Policy and governance in the water-energy-food nexus: A relational equity approach*. London: Routledge.

Koulouri, A., & Mouraviev, N. (2020b). Governance of the water-energy-food nexus: A relational equity approach. In A. Koulouri & N. Mouraviev (Eds.), *Policy and governance in the water-energy-food nexus: A relational equity approach*. London: Routledge.

Mouraviev, N., & Koulouri, A. (2017). Public-private partnerships as a policy tool for the sustainable utilisation of renewable energy sources: The case of

Kazakhstan. In N. Mouraviev & N. Kakabadse (Eds.), *Public-private partnerships in transitional nations: Policy, governance and praxis*. Newcastle upon Tyne: Cambridge Scholars Publishing.

Mouraviev, N., & Koulouri, A. (2018). Clean energy and governance challenges. *International Journal of Technology Intelligence and Planning, 12*(1), 1–5.

Mouraviev, N., & Koulouri, A. (Eds.). (2019a). *Energy security: Policy challenges and solutions for resource efficiency*. London: Palgrave Macmillan.

Mouraviev, N., & Koulouri, A. (2019b). Enabling green energy production: Implementing policy by using public-private collaboration. In N. Mouraviev & A. Koulouri (Eds.), *Energy security: Policy challenges and solutions for resource efficiency*. London: Palgrave Macmillan.

Mouraviev, N., & Koulouri, A. (2019c). Renewable energy and resource efficiency: Governance is key. In N. Mouraviev & A. Koulouri (Eds.), *Energy security: Policy challenges and solutions for resource efficiency*. London: Palgrave Macmillan.

Nazarbayev, N. (1997, October). *Prosperity, security and ever-growing welfare of all the Kazakhstanis—Strategy 2030*. Message of the President of the country to the people of Kazakhstan. Available via http://www.akorda.kz/en/official_documents/strategies_and_programs. Accessed 23 April 2020.

Nazarbayev, N. (2012, December). *New political course of the established state—Strategy 2050*. Address by the President of the Republic of Kazakhstan. Available via http://www.akorda.kz/en/official_documents/strategies_and_programs. Accessed 24 April 2020.

Schwab, K. (Ed.). (2019). *The global competitiveness report 2019*. Geneva: World Economic Forum. Available via http://www3.weforum.org/docs/WEF_TheGlobalCompetitivenessReport2019.pdf. Accessed 23 April 2020.

Trading Economics. (2020). *Kazakhstan exports by category*. Available via https://tradingeconomics.com/kazakhstan/exports-by-category. Accessed 24 April 2020.

Worldatlas. (2020). *The largest countries in the world*. Available via https://www.worldatlas.com/articles/the-largest-countries-in-the-world-the-biggest-nations-as-determined-by-total-land-area.html. Accessed 24 April 2020.

World Bank. (2019). *The World Bank in Kazakhstan—Country snapshot*. Available via http://pubdocs.worldbank.org/en/278551571374560680/Kazakhstan-Snapshot-Oct2019.pdf. Accessed 24 April 2020.

World Bank. (2020a). *Doing business 2020*. Washington, DC: World Bank. Available via https://openknowledge.worldbank.org/bitstream/handle/10986/32436/9781464814402.pdf. Accessed 23 April 2020.

World Bank. (2020b). *Kazakhstan—Overview*. Available via https://www.worldbank.org/en/country/kazakhstan/overview. Accessed 23 April 2020.

Worldometers. (2020). *Kazakhstan population (LIVE)*. Available via http://www.worldometers.info/world-population/kazakhstan-population/. Accessed 24 April 2020.

PART I

The Role of Key Sectors in Transforming the Nation

CHAPTER 2

Kazakhstan's Economy and the Oil Sector: A Developmental State or Resource Curse Model?

Serik Orazgaliyev and Slyamzhar Akhmetzharov

Introduction

Kazakhstan's oil and gas reserves are primarily (70%) located in the western part of the country where they are extracted from deposits in the Caspian Sea (Atyrau, West Kazakhstan, and Mangystau provinces) and the neighbouring provinces of Aktobe and Kyzylorda (KazMunaiGas 2020). Extraction of oil in the Caspian region started in the late-nineteenth century when the area currently occupied by modern-day Kazakhstan was part of the Russian empire. The rapid expansion of production was made possible with a joint investment from the Nobel brothers, the Rothschild family, and Shell company, who "made fortunes" from Caspian oil between 1880 and 1917 (Kinzer 1997). In the aftermath of the Bolshevik revolution of 1917 and the establishment of the Union of Soviet Socialist

S. Orazgaliyev (✉) · S. Akhmetzharov
Graduate School of Public Policy, Nazarbayev University,
Nur-Sultan, Kazakhstan
e-mail: serik.orazgaliyev@nu.edu.kz

A. Koulouri and N. Mouraviev (eds.),
Kazakhstan's Developmental Journey,
https://doi.org/10.1007/978-981-15-6899-2_2

Republics (USSR), the central government nationalised oil enterprises and forced out foreign investors. For more than 70 years, the Caspian region remained closed to foreign investment. With the collapse of the Soviet Union in 1991, the region re-opened its gates to oil multinationals.

The initial period following the disintegration of the Soviet Union proved to be difficult for the people and the governments of the post-Soviet states. Industries collapsed and both national and individual incomes decreased considerably. Agriculture, machine-building, and manufacturing, traditionally strong industries in most Soviet Republics, ceased production. In the light of these new circumstances, oil and gas resources were among the few valuable commodities that resource-rich post-Soviet states could export to international markets. This resulted in the development of the oil and gas sector becoming a matter of foremost importance for the newly independent Caspian states, including Kazakhstan.

Systemic issues related to access to capital, technology, and human resources have to date prevented Kazakhstan from attempting to develop oil and gas deposits on its own. The solution was found in attracting multinational oil companies to assist with the development of the nation's oil and gas resources. Among them, major oil multinationals, such as Chevron, ExxonMobil, Total, Eni, and Royal Dutch Shell, entered Kazakhstan's energy sector. Although multinational enterprises (MNEs) immediately contributed to the increase in oil production, the relations between Kazakhstan and oil MNEs have been far from harmonious. The initial excitement and euphoria that prevailed at the beginning of the partnership followed attempts to renegotiate agreements. MNE host country relations changed over time while the regulatory regime evolved responding to internal and external pressures.

Kazakhstan possesses about 30 billion barrels of proven oil reserves and 1 trillion cubic metres of proven natural gas reserves (BP 2019), which, in 2017, placed the country among the top 14 total energy producers and the top 12 oil producers in the world (Energy Information Administration [EIA] 2020). Moreover, the nation's economy is export-oriented with energy being the largest sector accounting for 21% of the country's gross domestic product (GDP), and approximately 62% of its exports in 2017 (World Bank 2019). Around three-quarters of Kazakhstan's oil is delivered to European markets, and about one-fifth is delivered to Asian markets, primarily to China (EIA 2015). Major energy consumers such as the United States, China, and the European Union have a genuine interest in cooperation regarding energy with Kazakhstan, as oil and gas supplies in the Caspian region could be perceived as an alternative to those in the Middle East. Moreover, with the development of stronger ties between Kazakhstan and China, including through projects related to China's One Belt and One Road Initiative[1] and the Kazakh-Chinese programme of industrial investment cooperation, it is expected that oil volumes exported to China will increase considerably in the next decade (Orazgaliyev 2020; Kazinform 2016a).

To this day, oil remains Kazakhstan's strategic asset, utilised by the government to achieve economic and geopolitical goals. Firstly, the oil sector drives the country's economy as it is the main export commodity, accounting for more than half of the country's exports, and hence its main source of budget revenue. Secondly, the oil sector plays a critical role in the realisation of the country's foreign policy agenda as energy cooperation with China, Russia, and the United States forms the basis of the country's relations with these major geopolitical powers (Orazgaliyev and Araral 2019).

This chapter primarily focuses on policy for the development of Kazakhstan's oil industry, analysing and discussing the implications of the country's dependence on oil revenues. In analysing the development and governance of the oil sector in Kazakhstan, the chapter draws on the elements of the developmental state concept (Johnson 1999), which is primarily used to explain the accelerated development of East Asian states—the so-called Asian Tigers—in the second half of the twentieth century (Woo-Cumings 1999). Since early independence, the Kazakhstani leadership used the East Asian path of development as a reference model. The country's First President Nursultan Nazarbayev, in his speeches, has praised highly the Singaporean experience of economic development (Michel 2015) and indicated that Kazakhstan follows a similar path (Araral 2016). The *Kazakhstan-2030* strategy, one of the key documents outlining the country's strategic direction, declares that the country is planning to transform itself into the "Central Asian Snow Leopard"

[1] This is an ambitious initiative aiming to establish land, along the ancient Silk Road through Eurasia, and maritime corridors to connect China with Europe and Africa. The estimated cost of all the projects under the One Belt and One Road Initiative umbrella is USD 575 billion (World Bank 2018). The Initiative is expected to support the construction of new infrastructure and to increase cooperation between the countries located along the routes ensuring enhanced policy coordination, greater financial integration, and expanding trade.

(Akorda 1997, section 2, para 3). It is therefore appropriate to evaluate the development and governance of Kazakhstan's oil sector through the lens of the developmental state concept, drawing from the experiences of the fast-developing East Asian countries.

This chapter is organised as follows. A brief argument on the suitability of the selected theoretical framework is provided, followed by an overview of the history of the oil sector in Kazakhstan during the pre- and post-independence periods. Foreign direct investment (FDI) in the oil sector and the latter's role in the country's economy are then discussed. Next, the government's steps to increase state participation in the industry and, hence, have a more direct input in the development of the sector in response to pressure from the electorate are considered. In particular, nationalisation policies and efforts to increase rent by amending existing legislation are examined. Further, Kazakhstan's governance model of the oil sector is critically discussed through the lens of the developmental state model and in comparison with examples of nations that have followed it. Finally, conclusions are drawn on the future policy directions to address the challenges Kazakhstan faces both in relation to the development of its oil sector, and more broadly its sustainable economic development.

Developmental State Model and the Southeast Asian Experience

The primary characteristics of the developmental state concept as identified by scholars are the presence of centralised government, and pervasive state intervention in the economy that is exemplified by the industrial policies employed and the existence of state-owned enterprises. It is often argued that these features provided the opportunity for countries such as South Korea, Taiwan, Singapore, and others to experience skyrocketing economic development from the 1960s to the1990s (e.g. Johnson 1982; Wade 1990; Thompson 1996).

One of the first scholars to attempt to explain the fast-growing tendency of East Asian countries was Chalmers Johnson (1982). He elaborated the developmental state concept to analyse, in particular, the case of post-war Japan. Johnson suggested that the role of the state in Japan significantly differed from the classical regulatory objectives. Instead, he pointed out, the Japanese government, in particular the Ministry of International Trade and Industry (MITI), was in charge of setting the direction for economic activities through extensive planning of industrial policies, incentivising desired outcomes, and stimulating technology transfer. The success of Japan in achieving consistently high levels of economic development relied not only on market forces, but also on the intervention of the state in the economy. Johnson (1999) argued that the developmental state concept incorporates the features of both rational market-type and centrally planned economies, and classified the developmental state as the "plan-rational state" (Johnson 1982, pp. 18–19).

The developmental state concept is characterised primarily by the presence of a strong state that is capable of setting and achieving economic objectives. In this regard, scholars point out that all fast-growing economies in the East Asia region were governed by strong, centralised and often authoritarian regimes in their initial period of development (Thompson 1996). For example, the skyrocketing economic development of South Korea was initiated under the regime of Major General Park Chung-Hee in the 1960s and 1970s (Minns 2001). The rapid economic growth of Taiwan between the 1950s and the 1980s took place under a single-party system (Wade 1990), whereas the Singaporean experience of fast economic development was under the long presidency of Lee Kuan Yew, from 1959 to 1990. Although post-World War II Japan was considered a democracy, its economic miracle has been associated with the implementation of the programme of the liberal-democratic party between 1955 and 1993.

Another characteristic of the developmental state concept is state intervention in the economy, through industrial policies targeted to strengthen particular sectors of the economy (Chang 1999). A notable example of this is the adherence of the South Korean political establishment to the goal of transforming the country's agrarian orientation into an industrial one, specialising in steel production (Wade 1990) supported by President Park's five-year plan, introduced in 1961, to advance iron and steel production. To achieve this, the government invested in infrastructure development (steel plants), and in enhancing the expertise of local specialists by sending Korean engineers for overseas training. In addition, South Korea used an import substitution strategy, accompanied by the strengthening of its export orientation. This involved the close monitoring by government officials of the prices and quality of domestic replacements to imported alternatives. Domestic firms tried to maintain government support by either ensuring price–quality trade-offs of their products for the internal market or by increasing their volume of exports. Similarly, in Taiwan, the industrial policy was associated with switching from labour-intensive to technology-intensive production, and accelerating the development of the high-tech sector with the inauguration of the Hsinchu Science Industrial Park in 1980 (Wong 1997).

Within the developmental state concept framework, government intervention in the economy may take place via state-owned enterprises (SOEs). For example, an SOE stood at the forefront of the South Korean steel production. The South Korean government established the Pohang Iron and Steel Company (POSCO) in 1968 (Chang 1999). By the late 1980s, POSCO had transformed into one of the world's most efficient steel producers. Similarly, SOEs played a crucial role in Taiwan's economic success, where the state-owned fertiliser industry provided employment as well as the opportunity to transfer surpluses to support industrial programmes in the country's less developed regions (Amsden 1979).

A key distinction between Kazakhstan and the abovementioned states is that Kazakhstan is both resource-rich and resource-dependent. Nonetheless, the developmental state concept is appropriate for the analysis of the development and governance of Kazakhstan's oil sector. This is because the country's leadership identifies with the developmental trajectory of these nations and has attempted to follow the East Asian developmental state model. It has, for example, been argued that the First President Nazarbayev employed the role of the "strong leader" in a similar fashion to leaders in South Korea, Taiwan, and Singapore (Araral 2016). Moreover, Kazakhstan has initiated a wide-ranging set of reforms and paid attention to modernisation strategies in a similar way to East Asian developmental countries. This chapter will therefore seek to explore the extent to which the Kazakhstani efforts have succeeded in replicating the success of the East Asian developmental model.

A Brief History of Oil Sector Development

Oil production in Kazakhstan started in 1899 when the Russian company Lehman & Co started developing the Karachungul deposit in the Emba region in western Kazakhstan. Discouraged by engineering challenges, the firm sold the project to British investors, who in turn established several private oil companies, including the Central Ural Caspian Company, North Caspian Oil Company, and Emba. All companies were British, with the exception of Emba, which had British, French, and German investors and the Nobel brothers among its shareholders (Maralbayev et al. 1999). By 1911, crude oil production reached commercial volumes, foreign investors launched oil exports from West Kazakhstan, and traders created "hype" about the "new Baku" and the enormous oil reserves in the Caspian Sea region. It all ended in 1917 following the Bolshevik revolution and the creation of the USSR; the new central government nationalised all oil companies, and forced foreign investors to leave.

Oil enterprises established in Kazakhstan throughout the 1950s and 1960s were actively producing during the Soviet period in fields in West Kazakhstan and in the country's central region. These fields were developed by state-owned enterprises Mangistaumunaigas, Uzenmunaigas, and Aktobemunaigas established in 1963, 1964, and 1967, respectively. At the time of independence, fields managed by the Aktobemunaigas enterprise produced annually about 17 million barrels of crude oil and had approximately 10,000 workers (JSC CNPC-AMG 2020).

The Soviet government developed not only the production facilities but also crude oil processing sites. Three relatively small oil refineries were built in Kazakhstan during the Soviet period: these include the Atyrau refinery (established 1945) near the Caspian Sea, the Pavlodar refinery (established 1976) near the Russian border, and the Shymkent refinery (established 1984) in South Kazakhstan (Maralbayev et al. 1999).

Nonetheless, during the Soviet era, Kazakhstan was not a major contributor in terms of oil and gas production, lagging behind the Soviet Republics of Azerbaijan and Turkmenistan, and the region of Siberia. The then Kazakh Socialist Republic produced only about 14% of the total Soviet production of oil and 2% of the total Soviet production of natural gas (Kozhokin 1998). By 1991, oil production in Kazakhstan was roughly at the same level as domestic consumption. Just before independence, in 1990, agriculture and livestock farming constituted over 40% of the country's GDP, whereas heavy industry was another large contributor to the economy. However, in the post-Soviet era, Kazakhstan has emerged as a leading oil producer, largely due to its vast oil and gas reserves developed by oil MNEs.

More recent oil discoveries in Kazakhstan include the fields of Tengiz (discovered in 1979), Karachaganak (discovered in 1979), and Kashagan (discovered in 2000). These are "supergiant" fields,[2] which hold vast

[2] Supergiant oil fields hold more than 5 billion barrels of recoverable oil, and supergiant gas fields hold more than 850 billion cubic meters of recoverable gas. (https://www.britannica.com/science/supergiant-oil-field and https://www.britannica.com/science/supergiant-natural-gas-field).

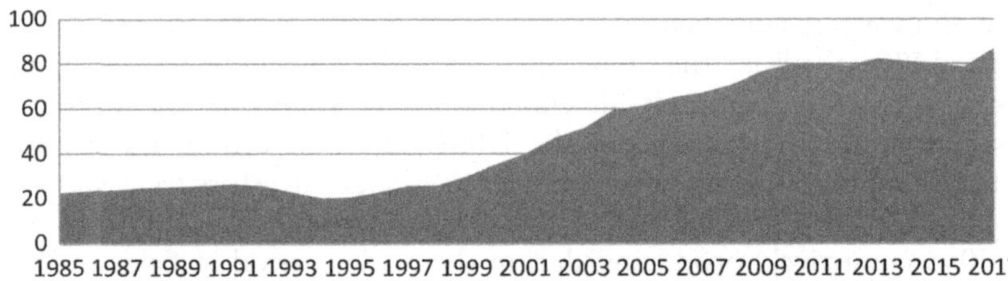

Fig. 2.1 Oil production in Kazakhstan, 1985–2017 (million tonnes per year) (*Source* Compiled by the authors from BP Statistical Review of World Energy 2019 [BP 2019])

oil and gas reserves. In 2016, approximately half of the oil produced in Kazakhstan was extracted from the Tengiz and the Karachaganak fields (EIA 2017). The successful development of these deposits enabled the increase of oil production from 26.6 million tonnes (182 million barrels) in 1991 to 87 million tonnes (0.6 billion barrels) in 2017 (Fig. 2.1). The Kashagan field started production in October 2016 and at full capacity will join the Tengiz and Karachaganak fields as one of the largest producing fields in the country (EIA 2017). The country's oil and gas production and exports are projected to increase with the start of production at the Kashagan field, and with the implementation of expansion projects in the Karachaganak and Tengiz fields. If current projects are realised, Kazakhstan's oil output will double, putting the country among the top 10 largest oil exporters in the world.

Foreign Investment in the Oil Sector

In the immediate years after the collapse of the Soviet Union and the emergence of the Republic of Kazakhstan in 1991, the country experienced economic difficulties caused by the disruption of production in key industries and in agriculture. The first four years of independence were characterised by negative growth, as the economy shrank by about 50% (Organisation for Economic Cooperation and Development [OECD] 2012). There was rising unemployment, decrease in household incomes, hyperinflation, and stagnation in most sectors of the economy. This was followed by an external shock in 1998 following the decline in commodity prices and the global financial crisis. To exit the recession, the government launched privatisation reforms in major sectors of the economy, including the country's oil sector. Kazakhstan's vast oil and gas reserves attracted "big name" oil MNEs including the US-based Chevron and ExxonMobil, and European firms, namely BG Group (UK) (which in 2015 was acquired by Royal Dutch Shell (UK–Netherlands)), Total SA (France), and Eni (Italy). Among non-Western MNEs operating in Kazakhstan are the Russian Lukoil and the Chinese National Petroleum Corporation (CNPC) (Table 2.1).

At the same time, international financial organisations (IFOs) supported Kazakhstan by providing expert advice and issuing loans. For example, with the assistance of the World Bank and the United States

Table 2.1 Kazakhstan's major oil and gas projects, proven reserves, and companies involved in their development

Field name	*Proven oil and gas reserves*	*Contract signed (contract term)*	*Shareholders (%, nationality)*
Tengiz	9 billion barrels of oil	1993 (40 years)	Chevron (50%, United States), Exxon Mobil (25%, United States), KazMunaiGas (KMG) (20%, KZ), and Lukoil (5%, RUS)
Karachaganak	2.5 billion barrels of oil and 1.2 trillion cubic metres (42 trillion cubic feet) of natural gas	1997 (40 years)	BG Group (29.25%, UK), Eni (29.25%, Italy), Chevron (18%, United States), Lukoil (13.5%, RUS), and KMG (10%, KZ)
Kashagan	13 billion barrels of oil	2001 (40 years)	Eni (16.81%, Italy), KMG (16.81%, KZ), Royal Dutch Shell (16.81%, UK–Netherlands), Total S.A. (16.81%, France), ExxonMobil (16.81%, United States), CNPC (8.4%, China), and Inpex (7.56%, Japan)

Source Compiled by the authors on the basis of PSA LLP (2019)

Agency for International Development (USAID), the government developed a mass privatisation programme, which included the privatisation of large enterprises in the oil sector. By 1997, Kazakhstan received 13 loans from the World Bank with a total value of approximately USD 1 billion (Jafar 2004). Between 1993 and 1998, the International Monetary Fund (IMF) provided loans totalling around USD 538 million to fund economic reforms and the government's economic programmes (IMF 2015). The funding was mainly used to build infrastructure projects, cover the budget deficit, and sustain the existing industrial capacity. Total net investments of the European Bank for Reconstruction and Development (EBRD) in Kazakhstan during the first two decades of independence amounted to about USD 7 billion, which was injected into 198 projects (EBRD 2019). IFOs had been particularly active in financing energy projects such as the construction of the Caspian Pipeline Consortium (CPC), which is currently the main export route of oil from the Tengiz field.

Considerable investment has been the main reason behind the expansion of oil production in Kazakhstan, and the reason for the increase in FDI inflows in the country. From the total volume of inward FDI around half constituted investment in the oil and gas sector (Olcott 2002). Between 1991 and 2019, the total volume of inward FDI amounted to USD 330 billion (Zhussupova 2019) with fluctuations of inflows between more than USD 10 million in 2006, and USD 22.5 billion in 2012, and with significantly higher levels of FDI inflows in the second decade after independence.

Fuelled by the increased prices in the global markets for crude oil, Kazakhstan's economy recovered and grew rapidly from 1999. In 2000, Kazakhstan's GDP grew by 9.6%, while during the next decade the growth rates have been among the highest in the world. Even after the economic crisis of 2008–2010, the country continued to be ranked in the top 10 fastest growing economies among the developing countries (World Bank 2013). This was followed by a period of unprecedented growth, with GDP per capita reaching a record USD 12,100 in 2012. However, this growth model eventually proved its unsustainability in 2015 when oil prices collapsed resulting in the country's economy going into recession. To compensate for the losses from the fall in oil prices, the currency was devalued and strict austerity measures were introduced. In August 2015, the National Bank of Kazakhstan introduced a freely floating exchange rate for the currency (the Kazakhstani Tenge, KZT), and from that point, the value of USD 1 increased from KZT 185 to KZT 330 by May 2016.

Kazakhstan's policy for attracting FDI is heavily influenced by the geopolitical environment and the country's foreign policy objectives. Kazakhstan officially declares adherence to a multi-vector approach in foreign policy, which is reflected in attempts to maintain good relations with the major geopolitical powers (Akorda 2014b). In the early 1990s, all of the country's oil export pipelines crossed the Russian territory, hence the necessity to balance Russian influence with the presence of US MNEs. Furthermore, international oil companies possess technical and financial capability to lead large-scale and complex energy projects. Starting from the late 1990s, China's cooperation with Kazakhstan in the energy sector has intensified, while the construction of the direct pipeline connecting the two countries eased exports to the East.

Dependence on Oil Revenues

In the aftermath of the Cold War, foreign investment was seen as a panacea for economic development in the post-Soviet countries. However, as it was demonstrated by the series of economic shocks caused by fluctuations in oil prices and global recessions, the FDI in the oil sector does not necessarily contribute to the sustainable development of the economy, but only provides short-term growth impetus. Moreover, under certain conditions, the expansion of the oil sector can negatively affect other important industries such as manufacturing and agriculture (Auty 2002), undermining long-term development.

Hence, a common problem among most resource-rich countries is the resource curse. Some analysts (Kutan and Wyzan 2005) identified symptoms of the Dutch disease in Kazakhstan, in that the expansion of the oil sector deterred the development of non-mineral sectors such as manufacturing and agriculture. Others (Egert and Leonard 2008), in contrast, argued that the development of the oil sector in Kazakhstan had not necessarily caused the contraction of other sectors; hence, they concluded that the economy is not affected by the Dutch disease. Similarly, the IMF (2004, p. 73) noted that "management of petroleum resources appears prudent, with few, if any, indicators of the resource curse". It should be noted that this report was published during a period of high oil prices, throughout 2001–2006, when the average GDP growth in Kazakhstan was 9%.

As oil revenues constitute the main source of GDP growth, the major concern for policymakers is the country's increasing dependence on this sector. Between 2000 and 2009, the GDP growth averaged 8.5%, while between 2009 and 2018 it averaged 5%. Real GDP per capita has more than doubled over the last two decades, from USD 4491 in 2000 to USD 11,165 in 2018. Moreover, the oil sector accounts for 17% of the country's GDP, and, in 2019, generated 44.3% revenue of the state budget (Forbes.kz 2020). Since 2001, the share of the oil industry in the gross exports of Kazakhstan has remained above 50%, reaching a historical high of 68% in 2014 (Ernst & Young 2013), whereas the increase of the share of the oil industry in the GDP—from 10.9% in 2001 to 25.2% in 2012—illustrates the growing dependence of the economy on mineral commodity exports.

Events in the near past have shown that contractions of the global economy can negatively affect Kazakhstan's economy (World Bank 2013). Firstly, during the global recession of 2008–2010 Kazakhstan's GDP fell by 1.2%, but the economy recovered well according to the World Bank (2013). Secondly, the oil crisis of 2014–2015 severely affected the economy and demonstrated its vulnerability to oil price volatility (Bershidsky 2015). The global price recovery allowed the increase of the share of oil exports to 64% in 2019 (Fig. 2.2), indicating once again the heavy reliance of the economy on the oil sector. The former head of the National Bank, Daniyar Akishev, pointed to the overarching role

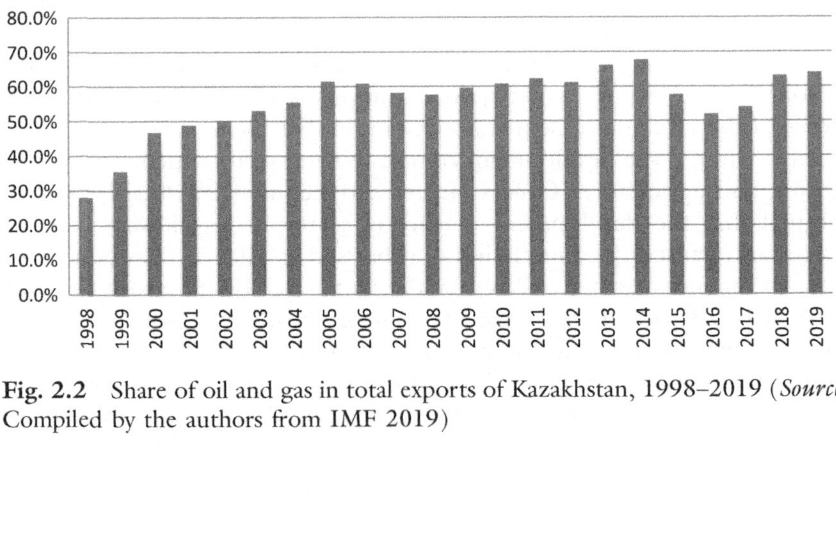

Fig. 2.2 Share of oil and gas in total exports of Kazakhstan, 1998–2019 (*Source* Compiled by the authors from IMF 2019)

of the oil sector in the country's exports and suggested that Kazakhstan's dependence on oil and gas is higher than, for example, that of Russia (Informburo.kz 2018).

The government has adopted policies to address the dependence on oil revenues with key strategic directions stated in the *Kazakhstan-2030* and the *Kazakhstan-2050* strategies (Akorda 1997, 2012). Furthermore, in 2001, the Samruk-Kazyna Sovereign Wealth Fund was established, designed to stabilise the economy by accumulating surpluses, and compensating losses during boom–bust cycles. It also serves as a savings fund for future generations. With the approval of the president, the government can process targeted transfers from the fund to the central or local budgets. These targeted transfers can be directed to major investment projects in the country or to stabilise the budget during periods of crisis. For example, transfers from the fund supported the domestic banking and construction sectors during the 2008–2009 financial crisis. Fund operations are managed by the National Bank, which reports that by the end of 2019, the fund accumulated USD 59.8 billion (National Bank of the Republic of Kazakhstan [NBRK] 2020).

In addition, the government has attempted to develop non-extractive sectors. Several projects have been launched throughout the country with the main goal of stimulating industrialisation through intensive investment in priority sectors. The Accelerated Industrial-Innovative Programme was introduced in 2010 to ensure "stable and well-balanced economic growth by means of diversification and improvement of its competitive ability" (Legal information system of the Ministry of Justice of the Republic of Kazakhstan 2010, section 1, para 4). Among the other policy measures undertaken by the government to ensure long-term economic upsurge are "the cluster economy" programme (Legal Information System of the Ministry of Justice of the Republic of Kazakhstan 2013), and the establishment of institutions such as the Development Bank of Kazakhstan (JSC DBRK 2020), the Investment Fund (JSC IFK 2020), the Innovation Fund[3] (JSC QazTech Ventures 2020), and the Export Credit Corporation[4] (JSC Kazakh Export 2020).

[3] Currently transformed into the JSC QazTech Ventures.

[4] Currently transformed into the JSC Kazakh Export.

Social Pressures for Nationalisation

The economic crisis of the 1990s forced the government to sell the nation's energy assets within a short time, resulting in the first-movers having an advantage when acquiring lucrative assets in the oil sector. A growing discontent about "giving away the country's wealth" became widespread in society. With the rise in oil prices, it became apparent that the Production Sharing Agreement (PSA) contracts, concluded with oil MNEs in the 1990s, failed to ensure fair rent distribution. The main concern of the general public was that the assets transferred to foreign investors during the 1990s were significantly undervalued to the benefit of the MNEs and bureaucrats (Ustiugov and Puzanov 1993). For example, Canadian company Hurricane Hydrocarbons purchased the Yuzhneftegaz enterprise in 1996 for USD 120 million and sold it in 2005 to CNPC for USD 4.20 billion. While opposition party activists questioned the legitimacy of the deal, government officials argued that the price was adequate given the financial hurdles that Kazakhstan experienced at the time.

Nonetheless, the criticism about "giving away the national wealth" was growing. The popular belief that foreign MNEs were taking a disproportionately large share of the rents persisted. The press openly criticised the government for expropriating capital earned from the privatisation of national assets, and social activists blamed it for what they saw as its inability to exercise control over oil MNEs. Inevitably, the question of fair distribution of energy export rents was politicised. Opposition parties used the popular discontent to challenge the government and propose countermeasures that proved popular with the electorate (Epokha 2004; Human Rights Watch 2004).

Public distrust was exacerbated by the lack of transparency in negotiations regarding the privatisation of the largest energy assets, particularly in the 1990s, which generated negative public sentiment regarding national oil asset privatisation. The detailed provisions of PSA contracts are not publicly released, and the only information available is that these agreements provided a range of guarantees for investors. These included contract stability mechanisms, provisions on international arbitration of disputes, acknowledgement of the priority of PSA contracts over national legislation, and guaranteed tax conditions. Kazakhstan also undertook obligations not to limit oil exports and refrain from imposing export duties (Adilov 2003).

In addition, the population of the nation's oil-rich regions and oil workers were largely unhappy with numerous issues including low wages, and the domination of MNEs in the oil sector. A particular issue was the disparity between the salaries and working conditions of foreign and local employees, with the former receiving significantly higher salaries and enjoying better working conditions compared to the latter for similar types of jobs (Sorbello 2019). This gave rise to wider labour movements in the oil-producing regions (Yessenova 2012). Negative sentiments led to several protests by local oilmen, which intensified after 2008 resulting in a violent clash between workers and the local police in the city of Zhanaozen in December 2011 (BBC News 2011).

The disparities in the compensation levels between foreign and local employees in the extractive industries have attracted the public's attention and remain on the government agenda. In November 2019, President Tokayev met with the CNPC leadership and urged them to "bring the wages of Kazakhstani employees to the level of foreign employees" (Kazinform 2019). Further, in December 2019, President Tokayev, during the second meeting of the National Council of Public Trust, announced the reduction of the quota for foreign workers by 40% from 49,000 to 29,000. He also emphasised that the Ministry of Labour would be addressing the problem of disparate wages among foreign and local employees (Akorda 2019).

Societal pressures have forced the government to take measures to increase its intervention in the oil industry. In response to the growing discontent in the society about "giving away the national wealth" and the unfair distribution of profits from the oil sector, the government decided to take action by amending the legislation regulating the industry. Starting from the early 2000s, Kazakhstan passed legislative reforms to ensure that the state budget gets a larger share of profits from oil exports. In addition, the government intervened in the sale of some of the private energy enterprises in Kazakhstan intending to make KMG an influential stakeholder in the industry (Orazgaliyev 2018). In the following section, the changes that the new legislation brought to actors in the oil sector are explored.

Changes in the Oil Sector Regulation

The new millennium brought the beginning of a shift of the bargaining powers from the MNEs to the government of Kazakhstan. Public discourse about the need to restore the balance of interests between the state and the MNEs had been pervasive. Chairing the session of the Foreign Investors Council, in 2001, the First President Nazarbayev signalled the change in the dynamics by saying that "the balance between the interests of the state and businesses" should be observed (Nazarbayev 2017, p. 138). Former Prime Minister and at the time head of the Kazakhoil national oil company (NOC) Nurlan Balgimbayev reflected the same rhetoric by stating in the same year, "[we] were called upon to ensure the balance of power in relations with foreign partners and to protect state interests" (Ostrowski 2010, p. 53).

Subsequently, the government attempted to revise legislation to strengthen its control over the sector. In 2004–2005 it passed amendments to the existing legislation governing the energy sector: the Law on Petroleum (Legal Information System of Regulatory Acts of the Republic of Kazakhstan [LIS of RA of the RoK] 2004a), the Subsoil Use Law (LIS of RA of the RoK 2004b), and the Law on Production Sharing Agreements (LIS of RA of the RoK 2005c).

Moreover, during the second decade after independence, Kazakhstan made efforts to strengthen the role of the country's national oil company, KMG, that was formed in 2002 through the merger of Kazakhoil and Oil and Gas Transportation. Legislative amendments allowed the state to exercise pre-emption rights on any oil assets put up for sale in Kazakhstan enabling KMG to purchase stakes in major energy projects in Kazakhstan. They further stipulated that KMG would have at least a 50% share in all new oilfield development projects. KMG was thus enabled to buy 33.3% of the PetroKazakhstan company, and 50% of the shares of MangistauMunaiGas (MMG) and Karazhanbasmunay, and to have 16.8% of the shares in the Kashagan, 20% in the Tengiz, and 10% in the Karachaganak fields.

Between 2004 and 2005, the Tax Code was amended (LIS of RA of the RoK 2004c, 2005b). Some of the changes specifically targeted oil companies. For example, a floating rent tax for oil exports was introduced, which was calculated based on current oil prices. At the same time, royalty payments increased from 0.5 to 2–6%, whereas the excess profit tax range was raised from 0–30 to 15–60%. Oil revenue tax was increased from 65 to 85%, although the new provision was applied only to new contracts concluded from January 2004 onwards. In 2009, the new Tax Code entered into force, which effectively replaced royalties with the natural resource extraction tax (LIS of RA of the RoK 2008). The law also introduced excess profit tax, which was due for payment after corporate tax payments. Although the new Tax Code abolished the PSA taxation regime, it recognised tax stability provisions in PSAs concluded prior to the introduction of the new legislation.

In the sphere of environmental protection, environmental organisations reported increased pollution levels and health problems of the citizens living in the oil production areas (Urbaniak et al. 2007). Kazakhstan moved to tighten regulations by introducing several changes to the existing legislation (LIS of RA of the RoK 2004d). Flaring of natural gas was banned and the requirement to move sulphur storages indoors was introduced. According to the 2010 version of the Law on Subsoil and Subsoil Use (LIS of RA of the RoK 2010), companies were obliged to utilise associated gas (i.e. gas associated with the processes of extractive industries) because "flaring of associated gas is a major resource waste and causes large emissions of greenhouse gases and air pollutants" (World Bank 2012). According to initial PSA contracts, the government could intervene and unilaterally cancel agreements in circumstances when oil operations threaten the health and safety of the local citizens. The government did therefore use ecological concerns as leverage to put pressure on MNEs.

For instance, the Ministry of Energy of the Republic of Kazakhstan (MoE) pressured PetroKazakhstan company by bringing lawsuits for price-fixing and violation of environmental regulation norms. Unable to settle disputes with the state agencies, PetroKazakhstan announced the sale of its shares to the CNPC and attempted to circumvent government plans to purchase a stake in the company. The MoE stated that the deal would not be approved, while some politicians called for the cancellation of the contract with PetroKazakhstan and the forced expropriation of its assets (Nurmakov 2010). However, the situation was resolved when the government passed an amendment to existing legislation, specifically designed to tackle it. The new petroleum law (LIS of RA of the RoK 2004a) granted the government a pre-emptive right to purchase not only energy assets in Kazakhstan but also equity stakes of energy companies operating in the country (LIS of RA of the RoK 2005a). Using this law, the MoE negotiated with the CNPC the transfer of 33% of shares of PetroKazakhstan and a 50% stake in the Shymkent oil refinery to KMG in 2006.

Another major dispute occurred in 2010 when the government accused the Karachaganak Petroleum Operating (KPO) for tax and environmental violations. In response, KPO brought an arbitration case against Kazakhstan, asserting that the claim of export duties was illegal.

The arbitration was later suspended as the parties started negotiations to settle the issue. Following protracted negotiations, the parties signed to an agreement with KMG acquiring 10% of shares in the project in exchange for USD 2 billion in cash, and USD 1 billion in a non-cash payment (Tengrinews 2011); the Samruk-Kazyna Sovereign Wealth Fund acted as a guarantor for the NOC in securing the payment.

Foreign investors perceived the increasing government intervention as the beginning of an aggressive nationalisation and heavily criticised the government. As a reaction to the changing regulatory framework, oil MNEs attempted to pressurise Kazakhstan directly and through their home governments. For example, in 2002, the ambassadors of the United States, Canada, and some of the European Union countries sent a joint letter to the government of Kazakhstan requesting to ease the pressure on foreign investors (Ostrowski 2010). In another case, in 2007, the Kazakhstani MoE warned the Italian company Eni regarding the consequences of delays in developing the Kashagan field. Shortly after, the then Italian Prime Minister, Romano Prodi, visited Astana, "to restore confidence but also to take on a number of commitments" (Guliyev and Akhrarkhodjaeva 2009, p. 3180). In response to these concerns, the government tried to reassure foreign investors that the state would not unilaterally change contract terms, stating that "resource nationalism is not the policy of Kazakhstan" (Asia_Society 2008).

In parallel, the MoE became more demanding in pursuing foreign investors to implement corporate responsibility and social programmes. Nonetheless, government attempts to persuade foreign investors to invest in local capacity-building programmes met with resistance. For instance, the subsoil legislation introduced in 1996 binds foreign investors to comply with local content requirements to buy at least 50% of materials from local producers (LIS of RA of the RoK 1996). The amendments to this legislation introduced in 1999 reinforced the requirement for operating companies to purchase local equipment, materials, and products, and use the services of local subcontracting companies (LIS of RA of the RoK 1999). The law stressed that the foreign firms' services should be used only if there were a lack of local suppliers and only with the approval of local authorities. It also stipulated that the regulator can unilaterally terminate contracts with firms that fail to comply with local content requirements.

Although the government urged MNEs to support the local content initiative, results have so far been poor. Despite very few MNEs fulfilling local content requirements, no contract terminations occurred. The weak enforcement of this aspect of the legislation was due to the lack of such requirements in the original PSAs with major oil MNEs (Chentsova and Brainina 2003) which included contract stability clauses enabling MNEs to avoid complying with some requirements posed by the new legislation. In the end, major oil MNEs successfully resisted state intervention in relation to local content requirements.

The government and oil companies are both interested in enhancing the effectiveness of the oil sector of the country. However, there is a radical difference between the two parties. The Kazakhstani government is inclined towards the principles of the developmental state concept, which prescribes a greater degree of state intervention and regulation with emphasis on protecting the national interests. In contrast, oil companies, involving both foreign and local producers, are interested in boosting their profits through market liberalisation and decreased government influence. These conflicting visions set obstacles for a productive collaboration between the two major stakeholders in the policymaking process.

Nonetheless, despite these conflicting visions and the changing regulatory regime, most of the oil MNEs continue to operate in an environment of increased state involvement. Moreover, it appears that the government will not discourage future FDI in the country's economy by engaging in direct nationalisation and expropriation of MNEs' assets. For example, in 2019, the government adopted a roadmap to double FDI inflows to the country by 2025 (Yergaliyeva 2019), and Prime Minster Mamin instructed the heads of central and regional governmental bodies to reinforce the attraction of foreign investment to the country, and search for opportunities to implement new large-scale industrial projects. In so doing, the state bodies are advised to cooperate with the Astana International Financial Centre (Prime Minister of the Republic of Kazakhstan 2019). Since access to capital and technology will remain crucial in developing the country's oil and gas deposits, Kazakhstan will seek to maintain the presence of MNEs. Moreover, considering the technological challenges in developing fields such as Tengiz and Kashagan, MNEs' know-how cannot be replaced outright by local expertise. Finally, the presence of MNEs will contribute to maintaining positive relations with key geopolitical actors in the region, thus ensuring the security of oil and gas exports.

Table 2.2 Changes to the regulations in the oil and gas sector

Year	*Laws/policies*	*Purpose*
1996 and 1999	Amendments to the Subsoil Law (LIS of RA of the RoK 1996, 1999)	Reinforcement of the requirement for operating companies to purchase local equipment, materials, and products, and the services of local subcontracting companies. Obligation of foreign investors to comply with local content requirements, and buy at least 50% of materials from local producers
2004 and 2005	Amendments to the Law on Petroleum (LIS of RA of the RoK 2004a, 2005a)	Amendments allowed the state to exercise pre-emption rights on any oil assets put up for sale in Kazakhstan, and stipulated that KMG should have a share of at least 50% in any new oilfield development project
2004	Amendments to the Subsoil Use Law (LIS of RA of the RoK 2004b)	
2005	Amendments to the Law on Production Sharing Agreements (LIS of RA of the RoK 2005c)	
Tax Code reforms		
2004 and 2005	Amendments to the Tax Code (LIS of RA of the RoK 2004c, 2005b)	• Introduction of the floating rent tax for oil exports, calculated on the basis of current oil prices • Royalty payments increased from 0.5 to 2–6% • Excess profit tax range raised from 0–30 to 15–60% • Oil revenue tax increased from 65 to 85%

(continued)

Table 2.2 provides an overview of the key legislative changes relating to the oil and gas sector.

Is the Oil Sector Governance Effective in Pursuing Kazakhstan's Development Goals?

Based on the preceding discussion of governance of the oil sector in Kazakhstan, this section will first explore which of the characteristics of the developmental state concept can be identified in the developmental

Table 2.2 (continued)

Year	*Laws/policies*	*Purpose*
2009	Tax Code	• Replacement of royalties with the natural resource extraction tax • Introduction of excess profit tax (due for payment after corporate tax payments)
Environmental protection		
2004	Amendments to the Law on Environmental protection (LIS of RA of the RoK 2004d)	Introduction of the ban on flaring of natural gas, and the requirement to move sulphur storages indoors
2010	Amendments On Subsoil and Subsoil Use (LIS of RA of the RoK 2010)	Introduction of the ban on flaring of associated gas, and the requirement to utilise it

Source Compiled by the authors

trajectory of Kazakhstan since it became an independent state in 1991. It will also consider the challenges government faces in governing the oil sector and more broadly ensuring the long-term sustainable development of the nation's economy.

Similarly to East Asian countries, Kazakhstan introduced various *industrial policies* aiming at developing particular sectors of the economy. The economic liberalisation reforms in the first years of independence were followed by active industrialisation efforts from the mid-2000s onwards. In this regard, as part of the *Kazakhstan-2030* strategy (Akorda 1997), the industrial policy for the years 2003–2015 was launched, which focused primarily on transport, health care, and education. In 2010, the government initiated the Programme for the Accelerated Industrial-Innovative Development for the period 2010–2014 (Legal Information System of the Ministry of Justice of the Republic of Kazakhstan 2010), which provided for broader access to financing for businesses, including loans with a subsidised interest rate. Furthermore, the government encouraged export orientation of production in agriculture, light manufacturing, mechanical engineering, pharmaceuticals, and other industries.

In December 2012, the First President Nazarbayev announced the achievement of the main goals set in *Kazakhstan-2030* (Akorda 1997).

In particular, he outlined that Kazakhstan had become a strong independent state with a sustainable democratisation process and multi-ethnic harmony, and the leading economy in the Commonwealth of Independent States (CIS).[5] Most importantly, he introduced a new long-term strategy *Kazakhstan-2050* (Akorda 2012), which identified ten global challenges that Kazakhstan is to face and indicated ways to meet them. Aiming to bring Kazakhstan into the league of the 30 most competitive economies in the world by 2050, under its framework, numerous new industrial policies were initiated. For instance, the industrial development strategy (2015–2019) focused on building technology-intensive sectors (Akorda 2014a), whereas the *Nurly Zhol* infrastructure development programme provided a USD 40 billion financial stimulus to various sectors including the energy infrastructure (Akorda 2015).

In addition, similarly to the early *political development* of East Asian countries, Kazakhstan followed a highly centralised government path at its early stages of development (Schatz 2009). The First President Nazarbayev headed the country for almost three decades, consistently winning elections by a significant margin. Political liberalisation has never been prioritised by the power holders in Kazakhstan (Sullivan 2018). Furthermore, the reformist agenda of the country has been driven by the motto "economy first, then politics" (Euronews 2010).

Within the framework of the developmental state concept, *government intervention in the economy* may take place through the creation of SOEs. For instance, in 1974, the Singaporean government created Temasek Holdings, which currently manages a number of major companies worth USD 308 billion (Temasek 2018). Similarly, the Samruk-Kazyna Sovereign Wealth Fund, which is controlled by the government, was created in Kazakhstan in 2008, and owns several major companies in the fields of transport, infrastructure, and extractive industries, including state-owned companies in the oil and gas sector, namely KMG and KazTransGas.

State intervention in market relations is vividly expressed in the regulation of procurement activities of subsoil users as shown in the local content policy (see relevant legislation in Table 2.2). The development of the oil industry could thus serve to stimulate small and medium-sized

[5] The CIS was formed following the dissolution of the USSR in 1991. Currently, it includes the following ex-Soviet nations: Armenia, Azerbaijan, Belarus, Kazakhstan, Kyrgyzstan, Moldova, Russia, Tajikistan, and Uzbekistan.

enterprises involved in the procurement processes. Although these businesses are directly linked to the demands of oil and gas companies, the local content policy ensures their growth by encouraging investment, and providing employment opportunities. In broader terms, local content requirements are designed to nurture infant industries. However, the results of such protectionist policies in Kazakhstan remain mixed. The policy produced results in the recruitment of local labour as oil and gas companies increasingly hire local personnel. However, due to the lack of technological capacity and the necessary human capital, it is difficult for oil and gas companies to fulfil the requirement of acquiring locally produced goods and services. In addition, extensive corruption in the procurement process does not provide a level playing field for all firms submitting procurement tenders (Kalyuzhnova and Belitski 2019).

When considering the oil sector in Kazakhstan, a *number of challenges* need to be addressed by policymakers.

Firstly, due to its landlocked location, the country has limited capacity in terms of oil export routes. The vast portion of Kazakhstani oil is exported through the CPC pipeline, which transits through Russia to the Black Sea. In 2018, about 75% of the total Kazakhstani exported oil left through this route (KazEnergy 2019). Further, the high cost of transporting oil by tankers to the Baku–Tbilisi–Ceyhan pipeline makes this route less attractive, while exports through the Kazakhstan–China pipeline have yet to reach their full capacity (Sputniknews 2017). As a result, the heavy dependence on a single route undermines the country's potential to expand its oil exports in the future.

Secondly, another challenge is the extensive government control of crude oil and oil products' prices in the domestic market. The government attempts to intervene by establishing price ceilings for petrol, and the anti-monopoly agency regularly issues fines to oil producers who do not comply with state-indicated petrol retail prices (Tengrinews 2018). With prices on the global market more competitive, oil producers are not motivated to supply crude oil to domestic refineries. This leads to periodic shortages of oil products in the domestic market (Satubaldina 2017). The government attempts to address this issue by imposing export and import quotas for crude oil and oil products; however, such measures are unsustainable in the long term due to Kazakhstan's membership in the World Trade Organisation and the Eurasian Economic Union.

Thirdly, the dominant position of the projects developing the supergiant fields of Tengiz, Karachaganak, and Kashagan undermines the potential for developing small and medium-sized oil companies. For example, in 2018, 78 small and medium-sized oil companies extracted only 10.5% of the total oil extracted in the country. This share has been relatively stable—at about 10% on average annually—over the last decade (KazEnergy 2019). However, the possibility for expansion of small and medium-sized companies is weakened by their vulnerability to external and domestic economic shocks. Moreover, these companies often lack the necessary infrastructure and technological capabilities, making them less attractive to foreign and domestic investors (KazEnergy 2019).

However, there are also two major challenges identified in the *Kazakhstan-2050* strategy: the exhaustibility of fossil fuel sources, particularly oil and gas, and the utilisation of the country's natural resource abundance to promote its long-term sustainable economic growth. Currently, oil and gas are not only primary energy sources, but also primary sources of budget revenue. While recognising this, the strategy acknowledges the need for a shift towards renewable energy sources and green technologies. A future challenge for Kazakhstan is thus developing green technologies and shifting the country's comparative advantage from the traditional to renewable energy sources. At the same time, it is necessary for the government to find ways to transform the focus of the country's economy, managing sustainably its reliance on fossil fuel sources before their depletion.

In view of this, it is worth noting that despite being one of the major oil exporters, Kazakhstan paradoxically faces difficulties in satisfying the domestic demand for petrol. *Kazakhstan-2050* sets the objective that by 2025 domestic producers "must fully provide their own market with fuels and lubricants of new environmental standards" (Akorda 2012, p. 14). This was followed by the modernisation of the three key refinery plants in the cities of Atyrau, Shymkent, and Pavlodar, which in turn led, in 2018, to increasing the total refinery output by 10.2% and petrol production by 17.2%.

The *Kazakhstan-2050* strategy states that the country should pursue the dual objective of remaining a major player in the global oil market while actively developing renewable energy sources such as solar and wind (Akorda 2012). The MoE, which is responsible for policymaking in the energy sector, outlined a long-term scenario, in which the growth of renewable energy should outperform that of traditional energy sources, such as oil and gas (Informburo.kz 2017). At the same time, the MoE indicated the intention to take advantage of the long-term increasing trends in global oil demand, that is predicted to continue growing until 2030. It therefore appears that in the foreseeable future the role of renewable energy in Kazakhstan will be supplementary.

During the 1970s to 2000s, Asian economies succeeded in moving from the third into the first world. These countries have successfully industrialised and transformed into high-income economies. In contrast, Kazakhstan, despite emulating the features of the developmental state concept, has failed to achieve similar outcomes being currently classified as an upper-middle-income economy (World Bank 2020). The limited success of implementing the developmental state model in Kazakhstan is related, to a large extent, to the failure of restructuring the economy. In *Kazakhstan-2050* major emphasis is placed on advancing new non-primary sector industries aimed for export, stating that gradual industrialisation will allow doubling the share of non-primary exports in total exports by 2025, and tripling by 2040.

The development and expansion of the oil sector has to date provided the opportunity for maintaining a relatively stable economic growth rate over the last two decades. It has, however, left the country's economy vulnerable to the volatility of oil markets, and the state dependent on oil for its budget revenues. Despite the government's efforts to restructure the economy, recent trends in the composition of total export dynamics demonstrate that oil and gas continue to constitute a vast portion of the country's exports.

Considering Kazakhstan's relatively unsuccessful (when compared to East Asian countries) implementation of the developmental state concept through the lens of the resource curse theory, it could be argued that the country's natural resource abundance was the reason behind it. The lack of natural resource wealth of Southeast Asian countries served as an enabler, motivating them to develop technologically intensive sectors. In contrast, since 1991, Kazakhstan's natural resource abundance set an economic development trajectory that heavily relied, and continues to rely, on the extraction and export of oil and gas. The dominant position of the oil sector in the Kazakhstani economy has therefore set obstacles for the development of other, non-primary sectors. As a result, regardless of the various strategic programmes supported by numerous policies, the government has had limited success in developing non-extractive sectors, such as machinery manufacturing and goods processing, which remain uncompetitive in the global market. The government therefore continues to struggle to increase the proportion of value-added goods in exports. This suggests that Kazakhstan's economy displays symptoms of the Dutch disease. Furthermore, periodic fluctuations in the oil and gas prices undermine the economic development of the country. Leading to regular currency devaluations, they forge an unattractive investment environment, discourage investment, and detrimentally affect the consumers' purchasing power. In the Kazakhstani context, the natural resource abundance undermines the effectiveness of the implementation of the developmental state concept, rendering it to a large extent a curse, rather than a blessing.

Conclusion

In the first years of transitioning from a planned to a market economy, Kazakhstan's policies aimed at attracting FDI in order to overturn economic stagnation. Following a period of privatisation reforms, government policies shifted to tighter regulation of oil industry with greater state intervention and attempts to increase rents from the sector. Societal pressures and calls for nationalisation pushed policymakers to attain a more visible role for the domestic national oil company, while maintaining the presence of MNEs. Despite the different perspectives of the government and of oil producers regarding the industry's governance, circumstances dictate a continued partnership. There is high potential for improved collaboration between the parties to promote sustainable development of the energy sector, ensuring domestic energy needs are met, expanding the development of renewable energy sources, introducing innovative sustainable energy technologies, and offering greater access to energy for citizens.

Almost three decades since Kazakhstan emerged as an independent state, the country's attempts to restructure the economy have not been as successful as first envisioned, and the economy remains heavily dependent on a single industry. As is often the case in resource-rich countries, the reliance on the energy sector renders other sectors, such as manufacturing and agriculture, uncompetitive, while the economy becomes fragile to external shocks due to price volatility on the global commodity markets (Sachs and Warner 2001; Auty 2002). Restructuring the economy, therefore, remains one of the key challenges for Kazakhstan in the coming years as the country faces an urgent need to shift its focus from the oil sector by implementing the necessary policy and governance reforms, and thus mitigating the economy's vulnerability to commodity market volatility.

Speaking at a plenary session of the IXth Astana Economic Forum in 2016 an esteemed economist Ha-Joon Chang delivered his policy recommendations and views on the future of the Kazakhstani economy (Kazinform 2016b). He asserted that Kazakhstan needs to develop its manufacturing industry since prices for oil and other primary products tend to decline in the long run. Chang argued elsewhere in his writings (Chang 2002, 2010) that industry protection and state intervention had been an important ingredient for the developmental success not only in Southeast Asia but also in Western Europe and the United States. Indeed, industrial development is the component, which is missing when one takes a closer analysis of Kazakhstan's economy, particularly, when comparing to the Southeast Asian states. All the other elements of the developmental state model can be found in Kazakhstan, including highly centralised governance, interventionist policies and the emergence of state-owned enterprises.

At the same time, recent developments might have a positive effect on the country's economy in the long term. Kazakhstan is a part of China's One Belt and One Road Initiative that is designed to promote connectivity and closer economic integration in the region and beyond. This can potentially transform the country into a trade and transport hub, linking Europe and Asia. Furthermore, Kazakhstan endeavours to transform itself into Central Asia's financial hub by establishing the Astana International Financial Centre (AIFC) to attract more foreign investment, open up the domestic banking sector, and make it easier for insurance companies and financial institutions to do business in the country. Overall, it is expected that closer economic integration in the region, and the improved investment climate will help to attract FDI, reinvigorate the economy, and help the government to ease its dependence on the oil sector for development.

References

Adilov, M. (2003). Commercial secrets of Tengiz courtyard [In Russian]. *Respublika, Kazakhstan's Weekly Newspaper*, October 24.

Akorda. (1997). *The strategy for development of the Republic of Kazakhstan—"Kazakhstan 2030"*. Available via https://www.akorda.kz/en/official_documents/strategies_and_programs. Accessed 10 April 2020.

Akorda. (2012). *Strategy "Kazakhstan 2050"—New political course of the established state*. Available via https://www.akorda.kz/en/addresses/addresses_of_president/address-by-the-president-of-the-republic-of-kazakhstan-leader-of-the-nation-nnazarbayev-strategy-kazakhstan-2050-new-political-course-of-the-established-state. Accessed 10 April 2020.

Akorda. (2014a). *Industrial development strategy of Kazakhstan for the period 2015–2019*. Available via https://www.akorda.kz/ru/official_documents/strategies_and_programs. Accessed 10 April 2020.

Akorda. (2014b). *On foreign policy concept of the Republic of Kazakhstan for 2014–2020*. Available via http://www.akorda.kz/ru/legal_acts/decrees/o-koncepcii-vneshnei-politiki-respubliki-kazahstan-na-2014-2020-gody. Accessed 2 May 2020.

Akorda. (2015). *State programme of infrastructure development "Nurly Zhol" for 2015–2019*. Available via https://www.akorda.kz/ru/official_documents/strategies_and_programs. Accessed 10 April 2020.

Akorda. (2019) *Speech of the president of the Republic of Kazakhstan K. Tokaev at the 2nd meeting of the National Council of Public Trust*. Available via http://www.akorda.kz/ru/speeches/internal_political_affairs/in_speeches_and_addresses/vystuplenie-glavy-gosudarstva-k-tokaeva-na-vtorom-zasedanii-nacionalnogo-soveta-obshchestvennogo-doveriya. Accessed 10 April 2020.

Amsden, A. H. (1979). Taiwan's economic history: A case of Etatisme and a challenge to dependency theory. *Modern China, 5*(3), 341–380. Available via https://doi.org/10.1177/009770047900500304. Accessed 28 April 2020.

Araral, E. (2016). Independent Kazakhstan following similar path as Singapore toward successful development. *The Astana Times, Opinions*, December 5. Available via https://astanatimes.com/2016/12/independent-kazakhstan-following-similar-path-as-singapore-toward-successful-development/. Accessed 28 April 2020.

Asia_Society. (2008). Statement of the Prime Minister Massimov. In *The Asia Society International Business Conference*, Astana, 13 June 2008. Available via http://asiasociety.org/pm-resource-nationalism-not-policy-kazakhstan. Accessed 28 April 2020.

Auty, R. M. (2002). *Sustaining development in mineral economies: The resource curse thesis*. London: Routledge.

BBC News. (2011, December 16). *Kazakh oil strike: 10 dead in Zhanaozen clashes*. BBC News Online. Available via https://www.bbc.com/news/world-asia-16221566. Accessed 2 April 2020.

Bershidsky, L. (2015). Kazakhstan is the latest oil-curse casualty. *Bloomberg Opinion*, August 20. Available via https://www.bloomberg.com/opinion/articles/2015-08-20/kazakhstan-is-the-latest-oil-curse-casualty. Accessed 2 April 2020.

BP. (2019). *BP statistical review of world energy 2019*. Available via https://www.bp.com/content/dam/bp/business-sites/en/global/corporate/pdfs/energy-economics/statistical-review/bp-stats-review-2019-full-report.pdf. Accessed 28 April 2020.

Chang, H.-J. (1999). The economic theory of the developmental state. In M. Woo-Cumings (Ed.), *The developmental state* (pp. 182–199). New York: Cornell University Press.

Chang, H.-J. (2002). *Kicking away the ladder: Development strategy in historical perspective*. London: Anthem Press.

Chang, H.-J. (2010). *Bad Samaritans: The myth of free trade and the secret history of capitalism*. New York: Bloomsbury.

Chentsova, O., & Brainina, N. (2003). New rules for the purchase of goods, work and services in the performance of petroleum operations in Kazakhstan. *Kazakhstan International Business Magazine 3*. Available via http://www.investkz.com/en/journals/32/254.html. Accessed 28 April 2020.

Egert, B., & Leonard, C. S. (2008). Dutch disease scare in Kazakhstan: Is it real? *Open Economies Review, 19*(2), 147–165. Available via https://www.researchgate.net/publication/226739822_Dutch_Disease_Scare_in_Kazakhstan_Is_it_real. Accessed 28 April 2020.

EIA. (2017). *US energy information administration, country analysis brief—Kazakhstan*. Available via http://www.ieee.es/en/Galerias/fichero/OtrasPublicaciones/Internacional/2017/EIA_Country_Aanlysis_Kazakhstan_10may2017.pdf. Accessed 28 April 2020.

EIA. (2020). *US energy information administration, total energy production*. Available via https://www.eia.gov/international/data/world. Accessed 27 April 2020.

Energy Information Administration (EIA). (2015). *US Energy Information Administration, International Energy data and analysis: China*. Available via https://www.eia.gov/international/content/analysis/countries_long/China/china.pdf. Accessed 28 April 2020.

Epokha. (2004, September 3). No title. *Kazakhstan's Weekly Newspaper, 35*(107).

Ernst & Young. (2013). *Global oil and gas tax guide*. Available via https://www.ey.com/Publication/vwLUAssets/2013_global_oil_and_gas_tax_guide/%24FILE/EY_Oil_and_Gas_2013.pdf. Accessed 28 April 2020.

Euronews. (2010 January 5). *Nazarbayev: "economy first, then politics"*. Interview with Nursultan Nazarbayev. Available via https://www.euronews.com/2010/01/15/nazarbayev-economy-first-then-politics. Accessed 6 April 2020.

European Bank for Reconstruction and Development (EBRD). (2019). *The EBRD in Kazakhstan*. Available via http://www.ebrd.com/kazakhstan.html. Accessed 28 April 2020.

Forbes.kz. (2020). *44% of state's budget in Kazakhstan is constituted from the oil and gas sector* [In Russian]. Available via https://forbes.kz//process/energetics/44_gosudarstvennogo_byudjeta_kazahstana_formiruet_neftegazovyiy_sektor/? Accessed 13 April 2020.

Guliyev, F., & Akhrarkhodjaeva, N. (2009). The Trans-Caspian energy route: Cronyism, competition and cooperation in Kazakh oil export. *Energy Policy, 37*(8), 3171–3182.

Human Rights Watch. (2004). *Political freedoms in Kazakhstan, D1603.* Available via https://www.refworld.org/docid/412eebaf4.html. Accessed 13 January 2020.

IMF. (2015). *Kazakhstan: Transactions with the fund from May 01, 1984 to March 31, 2020.* Available via https://www.imf.org/external/np/fin/tad/extrans1.aspx?memberKey1=535&endDate=2099%2D12%2D31. Accessed 28 April 2020.

IMF. (2019). *Republic of Kazakhstan: Article IV consultation-press release and staff report*. Available via https://www.imf.org/en/Publications/CR/Issues/2020/01/29/Republic-of-Kazakhstan-2019-Article-IV-Consultation-Press-Release-and-Staff-Report-49002. Accessed 28 April 2020.

Informburo.kz. (2017). *The energy strategy has been presented by the Ministry of Energy on WSEC-2017 in Astana* [In Kazakh]. Available via https://informburo.kz/stati/energeticheskuyu-strategiyu-kazahstana-predstavil-minenergo-na-wsec-2017-v-astane.html. Accessed 9 April 2020.

Informburo.kz. (2018). *Akishev: Kazakhstan depends on the export of oil and gas more than Russia* [In Kazakh]. Available via https://informburo.kz/novosti/akishev-kazahstan-zavisit-ot-nefti-i-metallov-bolshe-chem-rossiya.html. Accessed 28 April 2020.

International Monetary Fund (IMF). (2004). *Republic of Kazakhstan: Selected issues* (IMF Country Report No 04/362). Available via https://www.imf.org/external/pubs/ft/scr/2004/cr04362.pdf. Accessed 28 April 2020.

Jafar, M. (2004). Kazakhstan: Oil, politics and the new "great game". In S. Akiner (Ed.), *The Caspian: Politics, energy and security* (pp. 200–218). New York: Routledge.

Johnson, C. (1982). *MITI and the Japanese miracle: The growth of industrial policy, 1925–1975.* Stanford, CA: Stanford University Press.

Johnson, C. (1999). The developmental state: Odyssey of a concept. In M. Woo-Cumings (Ed.), *The developmental state* (pp. 32–60). New York: Cornell University Press. Available via https://www.academia.edu/3089669/The_developmental_state_Odyssey_of_a_concept. Accessed 28 April 2020.

JSC CNPC-AMG. (2020). *History of the JSC CNPC-Aktobemunaigas company.* Available via http://www.cnpc-amg.kz/?p=onas. Accessed 1 May 2020.

JSC DBRK. (2020). *About the Development Bank of Kazakhstan.* Available via https://kdb.kz/en/bank/about-us/. Accessed 28 April 2020.

JSC IFK. (2020). *History of the investment fund of Kazakhstan.* Available via https://www.ifk.kz/en/about/the-history-of-the-fund/. Accessed 28 April 2020.

JSC Kazakh Export. (2020). *History of the company*. Available via https://en.keg.kz/rubric/o_nas/history. Accessed 28 April 2020.

JSC QazTech Ventures. (2020). *History of the company*. Available via https://qaztech.vc/en/agency/history/. Accessed 28 April 2020.

Kalyuzhnova, Y., & Belitski, M. (2019). The impact of corruption and local content policy on firm performance: Evidence from Kazakhstan. *Resources Policy, 61*(January), 67–76. https://doi.org/10.1016/j.resourpol.2019.01.016.

KazEnergy. (2019). *The national energy report 2019*. Available via https://www.kazenergy.com/en/analyst/783/. Accessed 28 April 2020.

Kazinform. (2016a, September 21). *Kazakhstan-China to create 50 joint enterprises*. Kazinform, . Available via https://www.inform.kz/en/kazakhstan-china-to-create-50-joint-enterprises_a2951403. Accessed 27 April 2020.

Kazinform. (2016b). Nursultan Nazarbayev holds meeting within AEF. Retrieved from https://www.inform.kz/en/nursultan-nazarbayev-holds-meeting-within-aef_a2907882. Accessed 15 April 2020.

Kazinform. (2019, November 15). *President Tokayev holds meeting with CNPC board chairman*. Kazinform. Available via https://www.inform.kz/en/president-tokayev-holds-meeting-with-cnpc-board-chairman_a3585008. Accessed 2 April 2020.

KazMunaiGas. (2020). *Exploration production 2020*. Available via http://kmgep.kz/eng/about_kazakhstan/oil_and_gas_sector/. Accessed 27 April 2020.

Kinzer, S. (1997). Pipe dreams; A perilous new contest for the next oil prize. *New York Times*, September 21. Available via http://www.nytimes.com/1997/09/21/weekinreview/pipe-dreams-a-perilous-new-contest-for-the-next-oil-prize.html. Accessed 4 April 2020.

Kozhokin, E. M. (1998). *Kazakhstan: Realities and prospects of independent development* [In Russian]. Moscow: Russian Institute for Strategic Studies.

Kutan, A. M., & Wyzan, M. L. (2005). Explaining the real exchange rate in Kazakhstan, 1996–2003: Is Kazakhstan vulnerable to the Dutch disease? *Economic Systems, 29*(2), 242–255. Available via https://www.sciencedirect.com/science/article/abs/pii/S0939362505000440. Accessed 28 April 2020.

Legal Information System of Regulatory Acts of the Republic of Kazakhstan. (1996). *The subsoil and subsoil use law of the Republic of Kazakhstan* [In Russian]. Law No 2828 of 27 January 1996. Available via https://online.zakon.kz/Document/?doc_id=1004071. Accessed 28 April 2020.

Legal Information System of Regulatory Acts of the Republic of Kazakhstan. (1999). *Amendments to the subsoil law of the Republic of Kazakhstan* [In Russian]. Law No 467-1 of 11 August 1999. Available via https://online.zakon.kz/Document/?doc_id=1014277#pos=2;-153. Accessed 4 April 2020.

Legal Information System of Regulatory Acts of the Republic of Kazakhstan. (2004a). *Amendments to some legislative acts on oil operations in the Republic of Kazakhstan* [In Russian]. Law No 2-III of 1 December 2004. Available via https://online.zakon.kz/Document/?doc_id=1003801#pos=5;-153. Accessed 4 April 2020.

Legal Information System of Regulatory Acts of the Republic of Kazakhstan. (2004b). *Amendments to some legislative acts on subsoil use in the Republic of Kazakhstan* [In Russian]. Law No 2-III of 1 December 2004. Available via https://online.zakon.kz/Document/?doc_id=1052043. Accessed 4 April 2020.

Legal Information System of Regulatory Acts of the Republic of Kazakhstan. (2004c). *Amendments to some legislative acts on tax matters in the Republic of Kazakhstan* [In Russian]. Law No 11-III 13 December 2004. Available via https://online.zakon.kz/document/?doc_id=1052291#pos=0;116. Accessed 4 April 2020.

Legal Information System of Regulatory Acts of the Republic of Kazakhstan. (2004d). *Amendments to the law of the Republic of Kazakhstan "on environmental protection"* [In Russian]. Law No 8-III of 9 December 2004. Available via https://online.zakon.kz/document/?doc_id=1052192. Accessed 4 April 2020.

Legal Information System of Regulatory Acts of the Republic of Kazakhstan. (2005a). *Amendments to some legislative acts on oil operations in the Republic of Kazakhstan* [In Russian]. Law No 79-III of 14 October 2005. Available via https://online.zakon.kz/document/?doc_id=30028439#pos=0;0. Accessed 4 April 2020.

Legal Information System of Regulatory Acts of the Republic of Kazakhstan. (2005b). *Amendments to some legislative acts on tax matters in the Republic of Kazakhstan* [In Russian]. Law No 89-III of 22 November 2005. Available via https://online.zakon.kz/document/?doc_id=30034752#sub_id=80. Accessed 4 April 2020.

Legal Information System of Regulatory Acts of the Republic of Kazakhstan. (2005c). *The law on production sharing agreements of the Republic of Kazakhstan* [In Russian]. Law No 68-III of 8 July 2005. Available via https://online.zakon.kz/Document/?doc_id=30015925#pos=2;-155. Accessed 4 April 2020.

Legal Information System of Regulatory Acts of the Republic of Kazakhstan (2008). *Code of the Republic of Kazakhstan about taxes and other obligatory payments to the budget* [In Russian]. № 99-IV of 10 December 2008. Available via https://online.zakon.kz/document/?doc_id=30366217#pos=4;-106. Accessed 1 May 2020.

Legal Information System of Regulatory Acts of the Republic of Kazakhstan. (2010). *The law on subsoil and subsoil use in the Republic of Kazakhstan* [In Russian]. Law No 291-IV of 24 June 2010. Available via https://www.wto.

org/english/thewto_e/acc_e/kaz_e/WTACCKAZ69_LEG_3.pdf. Accessed 7 May 2020.

Legal Information System of the Ministry of Justice of the Republic of Kazakhstan. (2010). *The accelerated industrial innovative programme of the Republic of Kazakhstan for 2010–2014* [In Russian]. Law No 958 of 19 March 2010. Available via http://adilet.zan.kz/rus/docs/U100000958_/links. Accessed 4 April 2020.

Legal Information System of the Ministry of Justice of the Republic of Kazakhstan. (2013). *Concept for the formation of promising national clusters of the Republic of Kazakhstan until 2020* [In Russian]. Law No 1092 of 11 October 2013. Available via http://adilet.zan.kz/rus/docs/P1300001092. Accessed 4 April 2020.

Maralbayev, N., Atchibayev, A., & Adayaev, Z. (1999). *Petroleum encyclopedia of Kazakhstan*. Astana: Zholdas & Co.

Michel, C. (2015). Lee Kuan Yew's Singapore as a model for Kazakhstan. *The Diplomat*, March 30. Available via https://thediplomat.com/2015/03/lee-kuan-yews-singapore-as-a-model-for-kazakhstan/. Accessed 1 April 2020.

Minns, J. (2001). Of miracles and models: The rise and decline of the developmental state in South Korea. *Third World Quarterly, 22*(6), 1025–1043. Available via https://doi.org/10.1080/01436590120099777. Accessed 28 April 2020.

National Bank of the Republic of Kazakhstan (NBRK). (2020). *International reserves and assets of the National Oil Fund of the Republic of Kazakhstan* [In Russian]. Available via https://nationalbank.kz/?docid=285&switch=russian. Accessed 28 April 2020.

Nazarbayev, N. (2017). *The era of independence* [In Russian]. Astana.

Nurmakov, A. (2010). Resource nationalism in Kazakhstan's petroleum sector: Curse or blessing? In I. Overland, H. Kjaernet, & A. Kendall-Taylor (Eds.), *Caspian energy politics: Azerbaijan, Kazakhstan and Turkmenistan* (pp. 20–37). London: Routledge.

Olcott, M. (2002). *Kazakhstan: Unfulfilled promise*. Washington, DC: Carnegie Endowment for International Peace.

Orazgaliyev, S. (2018). State intervention in Kazakhstan's energy sector: Nationalisation or participation? *Journal of Eurasian Studies, 9*(2), 143–151. Available via https://www.sciencedirect.com/science/article/pii/S1879366518300125?via%3Dihub. Accessed 28 April 2020.

Orazgaliyev, S. (2020). The role of home governments in entry bargaining of multinationals: China's investment in Central Asia's natural resource sector. *Thunderbird International Business Review, 62*(3), 249–262. https://doi.org/10.1002/tie.22128. Accessed 28 April 2020.

Orazgaliyev, S., & Araral, E. (2019). Conflict and cooperation in global commons: Theory and evidence from the Caspian Sea. *International Journal of the Commons, 13*(2), 962–976. https://doi.org/10.5334/ijc.914. Accessed 28 April 2020.

Organisation for Economic Cooperation and Development (OECD). (2012). *OECD investment policy reviews: Kazakhstan 2012, OECD investment policy reviews*. Paris: OECD Publishing. Available via https://doi.org/10.1787/9789264121812-en. Accessed 28 April 2020.

Ostrowski, W. (2010). *Politics and oil in Kazakhstan*. Abingdon: Routledge.

Prime Minister of the Republic of Kazakhstan. (2019). *Askar Mamin instructs regions to cooperate with AIFC to attract investment*. Official information source of the Prime Minister of the Republic of Kazakhstan. Available via https://primeminister.kz/en/news/askar-mamin-poruchil-regionam-sotrudnichat-s-mfca-dlya-privlecheniya-investiciy. Accessed 6 April 2020.

PSA LLP. (2019). *North Caspian project* [In Russian]. Available via http://psa.kz/proekty Accessed 28 April 2020.

Sachs, J. D., & Warner, A. M. (2001). The curse of natural resources. *European Economic Review, 45*(4–6), 827–838. Available via https://www.sciencedirect.com/science/article/pii/S0014292101001258. Accessed 28 April 2020.

Satubaldina, A. (2017). Energy minister explains recent petrol price hike in Kazakhstan, gets reprimanded. *The Astana Times, Business*, October 11. Available via https://astanatimes.com/2017/10/energy-minister-explains-recent-petrol-price-hike-in-kazakhstan-gets-reprimanded/. Accessed 14 April 2020.

Schatz, E. (2009). The soft authoritarian tool kit: Agenda-setting power in Kazakhstan and Kyrgyzstan. *Comparative Politics, 41*(2), 203–222. Available via https://doi.org/10.5129/001041509x12911362972034. Accessed 28 April 2020.

Sorbello, P. (2019). Mass brawl at Kazakh oil field unveils labor dissatisfaction. *The Diplomat*, July 2. Available via https://thediplomat.com/2019/07/mass-brawl-at-kazakh-oil-field-unveils-labor-dissatisfaction/. Accessed 2 April 2020.

Sputniknews. (2017, April 5). *Transit of oil from Kashagan to Baku became too expensive* [In Russian]. Sputniknews. Available via https://az.sputniknews.ru/economy/20170405/409670769/tranzit-neft-kazahstan-baku-nevygodno.html. Accessed 13 April 2020.

Sullivan, C. J. (2018). Kazakhstan at a crossroads. *National Bureau of Asian Research. Asia Policy, 13*(2), 121–136. https://doi.org/10.1353/asp.2018.0028. Accessed 28 April 2020.

Temasek. (2018, July 10). *Temasek review 2018: Record net portfolio of S$308 billion*. Available via https://www.temasek.com.sg/en/news-and-views/news-room/news/2018/temasek-review-2018–record-net-portfolio-value-of-s-308-billion. Accessed 11 April 2020.

Tengrinews. (2011). *Kazakhstan completes acquisition of a 10% stake in Karachaganak*. 28 June 2012. Available via https://en.tengrinews.kz/companies/

Kazakhstan-completes-acquisition-of-a-10-stake-in-11144/. Accessed 28 April 2020.

Tengrinews. (2018, September 12). *Petrol stations have been penalized for the high prices on fuel* [In Russian]. Available via https://tengrinews.kz/kazakhstan_news/azs-oshtrafovali-za-zavyishennyie-tsenyi-na-benzin-v-almatyi-353405/. Accessed 14 April 2020.

Thompson, M. R. (1996). Late industrialisers, late democratisers: Developmental states in the Asia-Pacific. *Third World Quarterly, 17*(4), 625–648. https://doi.org/10.1080/01436599615290. Accessed 28 April 2020.

Urbaniak, D., Gerebizza, E., Wasse, G., & Kochladze, M. (2007, December 1). *Kashagan oil field development, Kazakhstan.* Friends of the Earth Europe. Available via http://www.foeeurope.org/publications/2007/KashaganReport.pdf. Accessed 28 April 2020.

Ustiugov, M., & Puzanov, O. (1993). Big oil moves in. *Bulletin of the Atomic Scientists, 49*(8), 44–47. Available via https://www.tandfonline.com/doi/pdf/10.1080/00963402.1993.11456401. Accessed 28 April 2020.

Wade, R. (1990). *Governing the market: Economic theory and the role of government in East Asian industrialisation.* Princeton: Princeton University Press.

Wong, J. (1997). The adaptive developmental state in East Asia. *Journal of East Asian Studies, 4*(3), 345–362. Available via https://www.cambridge.org/core/journals/journal-of-east-asian-studies/article/adaptive-developmental-state-in-east-asia/772491D441064B9620CA2F973C397181. Accessed 28 April 2020.

Woo-Cumings, M. (Ed.). (1999). *The developmental state.* New York: Cornell University Press.

World Bank. (2012). *Associated gas utilization via miniGTL.* World Bank. Available via https://openknowledge.worldbank.org/handle/10986/21976. Accessed 28 April 2020.

World Bank. (2013). *Kazakhstan—On the crest of the oil wave: Kazakhstan economic update, Spring 2013.* World Bank. Available via https://openknowledge.worldbank.org/handle/10986/22126. Accessed 28 April 2020.

World Bank. (2018, March 29). *Belt and Road Initiative.* Available via https://www.worldbank.org/en/topic/regional-integration/brief/belt-and-road-initiative. Accessed 28 April 2020.

World Bank. (2019). *The World Bank DataBank.* Available via https://databank.worldbank.org/reports.aspx?source=2&country=KAZ. Accessed 28 April 2020.

World Bank. (2020). *The World Bank in Kazakhstan—Overview.* Available via https://www.worldbank.org/en/country/kazakhstan/overview. Accessed 6 April 2020.

Yergaliyeva, A. (2019, July 13). Kazakh government adopts roadmap to double foreign direct investment by 2025. *The Astana Times, Business*. Available via https://astanatimes.com/2019/07/kazakh-government-adopts-roadmap-to-double-foreign-direct-investment-by-2025/. Accessed 6 April 2020.

Yessenova, S. (2012). The Tengiz oil enclave: Labor, business, and the state. *POLAR-Political and Legal Anthropology Review, 35*(1), 94–114. https://doi.org/10.1111/j.1555-2934.2012.01181.x. Accessed 28 April 2020.

Zhussupova, D. (2019, September 11). Kazakhstan attracts $330 billion FDI since 1991. *The Astana Times, Nation*. Available via https://astanatimes.com/2019/09/kazakhstan-attracts-330-billion-fdi-since-1991/. Accessed 28 April 2020.

CHAPTER 3

Evolution of Agricultural Policy in Kazakhstan

Aziza Baubekova, Anastasia Tikhonova, and Anastasia Kvasha

Introduction: The Country's Historical Path

Traditional pastoral and nomadic agriculture prevailed in Kazakhstan until the middle of the nineteenth century. With the growing influence of neighbouring Russia, Slav settlers started introducing sedentary farming and grain crops in the south-eastern, rain-fed territories. In the 1860s, after the Russian Empire expanded to Central Asia, cotton became the main crop cultivated in southern Kazakhstan, particularly in the irrigated regions of the Syrdarya Valley (Feher and Fieldsend 2019).

During the Soviet era, Kazakhstan was a major agricultural producer, supplying other Soviet republics with crop and livestock. Agriculture accounted for a significant share (about 30%) of the republic's gross

A. Baubekova (✉)
Center for Preparatory Studies, Nazarbayev University, Nur-Sultan, Kazakhstan
e-mail: aziza.baubekova@nu.edu.kz

A. Tikhonova
University of Yangon, Yangon, Myanmar

A. Kvasha
Central European University, Budapest, Hungary

A. Koulouri and N. Mouraviev (eds.),
Kazakhstan's Developmental Journey,
https://doi.org/10.1007/978-981-15-6899-2_3

domestic product (GDP). In the 1980s, Kazakhstan produced nearly 7% of the total Soviet agricultural gross product, and approximately 13% of the grain (Gaisina 2011). Around this period, two notable policy decisions took place.

The first policy decision was the enforced *collectivisation* of 1928–1931, which resulted in a huge reduction of livestock (the total herd size dropped by 80%) and devastating famine in the 1930s (around 1.5 million deaths). Implementing this policy, carried out across the entire Soviet Union, individual rural households were forced to consolidate into collective farms. Three different types of such agricultural enterprises were introduced in Kazakhstan: (i) a commune (a fully communalised farm with all land, animals, and capital in collective use); (ii) an artel (land, animals, and capital were still owned by the farm, but individuals were allowed to have a small private plot and a few animals); and (iii) an Association for Joint Cultivation of Land (land in common use, but individual ownership of animals permitted) (Olcott 1981).

The second policy decision was the *Virgin Lands programme* (1954–1960) introduced in the northern steppe lands of Kazakhstan. This decision brought around 25 million hectares of new lands into cultivation (over 60% of the current area of arable land) and made the republic the largest producer of wheat and barley in the Soviet Union. However, due to unfavourable climatic conditions causing unstable yields, around 30% of these lands turned out to be unsuitable for long-term cultivation and parts were later abandoned after the Soviet Union collapsed in 1991 (Feher and Fieldsend 2019).

The agricultural sector used to receive substantial support during the Soviet era. The state provided farmers with inputs (fuel, fertilisers, quality seeds) at much lower prices compared to those in the world market. Agricultural loans were given based on production plans, rather than on consideration of the borrower's creditworthiness. As a result, the loan repayment ratio was very low, the government regularly wrote off the debts, and farmers used to consider government lending more as grants, rather than actual loans (Gaisina 2011).

Nowadays, agriculture accounts for nearly 6% of Kazakhstan's economic production (Feher and Fieldsend 2019), with around 18% of the working population employed in the sector, while 43% of the total workforce reside in rural areas (Food and Agriculture Organization [FAO] 2017). Over the past few years, the sector has been growing on average at a rate of 4.4% per year (Tokbergenova et al. 2018). While only one-third of the country's land is currently used for agricultural production, around three-quarters of the total territory is considered suitable for such activities (Feher and Fieldsend 2019). Furthermore, the area of land suitable for grazing is one of the largest in the world, amounting to 181 million hectares (Tokbergenova et al. 2018).

Farmers mainly cultivate wheat, barley, oilseeds, cotton, rice, sugar beet, tobacco, forage crops, fruit, and vegetables. Animal husbandry focuses on sheep and cattle, producing dairy goods, leather, meat, and wool. Globally, Kazakhstan ranks 12th in terms of wheat production and 6th in wheat exports (Feher and Fieldsend 2019). In 2015, the agricultural sector was responsible for 4.8% of the total volume of the country's exports, represented largely by wheat and livestock products (Tokbergenova et al. 2018). The share of agricultural goods in the republic's total imports was around 11% in 2017, with meat, dairy products, fruit, and vegetables being the main imported items (Tokbergenova et al. 2018).

As a consequence of policy decisions made during the Soviet time and lack of investment in the sector during the first decade of Kazakhstan's independence, agricultural lands incurred substantial humus losses (Organisation for Economic Co-operation and Development [OECD] 2013). Nowadays, large areas of pastures are overgrazed and degraded (Tokbergenova et al. 2018), while water resources are limited and unevenly distributed across the country. Despite this, the agricultural sector remains the major water consumer, with the largest share used for irrigation (Baubekova and Kvasha 2019). The relatively harsh climate, which ranges from strongly continental in the north to arid with irregular rainfall in the centre, is further exacerbated by climate change, potentially leading to more frequent floods and droughts, and harming crop and livestock production activities in the country (OECD 2013).

This chapter explores the progression of Kazakhstan's agricultural sector and related policies since the nation gained independence in 1991. The chapter begins by examining different stages in the sector's development, starting from the early years of independence, throughout the years of the economic crisis of the 1990s, and the beginning of land and financial reforms. The following section focuses on the agricultural sector's recovery that started around the year 2000, providing an overview of the policy instruments and agreements introduced at that time. Finally, most recent developments and trends in agricultural policy are discussed, particularly the aspect of climate change adaptation, as well as possible future steps towards sustainability.

Early Years of Independence: Disruption of Economic Ties and a Significant Decrease in Agricultural Output

After the dissolution of the Soviet Union in 1991 and the subsequent price liberalisation, farmers could no longer benefit from paying below global prices for inputs and earning above world-level prices for their output. Supply chains were disrupted. In addition, during the hyperinflation of 1992–1994, input prices increased twice as much as output prices (Feher and Fieldsend 2019). In the 1990s, many areas of agricultural policy were largely neglected by the government as there were several more pressing issues at that time, such as macroeconomic stabilisation and transition to a market economy, along with a corresponding need for the creation of a basic legal framework (Feher and Fieldsend 2019).

Subsidies for the sector were significantly reduced from 10% of the GDP in 1991 to 2% in 1993 and were totally abolished in 1995. As a result, the volume of agricultural production decreased by about 55% between 1991 and 1998 (Gaisina 2011). In 1993, the deficit of cash inflow resulted in bartering becoming the main tool of trade from 1993 to 1998. Many remote farms lacked essential resources and went bankrupt. Access to credit at this time became critical but was hampered by high interest rates and a lack of collateral (Gaisina 2011).

As a result, in the second half of the 1990s, the agricultural sector in Kazakhstan entered a deep crisis (Feher and Fieldsend 2019). Between 1990 and 2001, the annual growth rate of agriculture value added (i.e. the net output of the agricultural sector after subtracting intermediate inputs) was negative (−3.2%). Annual grain production decreased from 30 million tons in 1992 to 6.5 million tons in 1998. Large-scale livestock farming almost disappeared, as livestock production shifted to small households, and meat, milk, and eggs became essentially non-traded goods (Petrick and Pomfret 2016). This basically meant that the prevailing type of agriculture in the country was subsistence farming, while supply of urban areas with food items relied principally on imports. The share of agriculture in GDP fell dramatically from 33.9% in 1990 to only 8.4% in 1998 (Government of Kazakhstan 1998).

Early Stages of Land Reform

Land reform plays one of the most important and defining roles in the agricultural policy transformation in Kazakhstan. It began in 1991 and had an uneven character (OECD 2013). Even now, private ownership of land in Kazakhstan remains a pressing issue, due to the large share of the country's population living in rural areas and a continuing reliance on its own agricultural production and land. The reform's first stage took place in the period 1991–1995 and focused on the reorganisation of state farms (*sovkhoz*—state-owned and state-operated agricultural enterprises) and collective farms (*kolkhoz*—cooperative enterprises operated by workers who belonged to a formed group, on state-owned land)—the dominant type of farms during the Soviet time. The reform aimed at transforming these large enterprises into smaller corporate farms and established individual usage rights to land, without changing the land ownership (Gaisina 2011). The latter remained in the hands of the state.

The 1995 Presidential Decree "On Land" stated that the land plots of new agricultural enterprises must be divided among the members of former production cooperatives, which included not only agricultural workers but also those involved in non-agricultural activities and pensioners (Presidential Decree 1995). These members received land plot certificates entitling them to a share of a farm's land, which, in fact, marked the beginning of land privatisation in Kazakhstan. However, the land distribution was made only in the form of usufruct[1] (Spoor 1999), while issued certificates did not even define the location and borders of a particular land plot. Physically the land remained in the hands of the production cooperative and was distributed "in-kind" only in the case when a certificate-holder would leave the enterprise to start private farming (Gaisina 2011). The Presidential Decree allowed individuals and legal entities to get a long-term lease of land plots for 99 years (Presidential Decree 1995). In 2001, the new Law on Land reduced this time to 49 years (Law of the Republic of Kazakhstan 2001b).

The second stage of land reform took place between 1996 and 2003 and was marked by the introduction of private land ownership on household plots in 1998. At the same time, the rest of the agricultural land remained under tenancy agreements until 2003, when the new Land

[1] Usufruct—the right to personal use and possession, excluding, however, the right to sell, damage, or destroy the property (Skryabin 2011).

Code was signed into law (Land Code 2003). However, the reform progressed slowly and unevenly, with the real situation on the ground often lagging behind the legislation (OECD 2013). New policies were not able to reinforce tenure rights or facilitate land market development. Prior to 2003, the only way to possess agricultural land was through a lease (of up to 49 years, with an option to sub-lease) (Gaisina 2011). While the new law would introduce private ownership of the agricultural land, which was a much-needed step towards a market-oriented economy, many politicians opposed the new Land Code. They argued that only the wealthy would benefit from the new legislation and the land would not be distributed fairly. Some of the Code's restrictive provisions (on ownership and leasing) also affected the functioning of the land market (Deshpande 2006). Years after the introduction of the new law, the agricultural land market remained underdeveloped as the proportion of privately owned land continued to be small due to the high prices of the state land and unclear regulations (Gaisina 2011).

Farm Restructuring

With the aim of transforming the principles of land ownership, this reform has led to significant changes in the structure of the country's farming sector. The former state-owned farms were dissolved and new collective farm entities were created in their place. The number of state farms dropped from 2500 in 1991, to only 89 in 2000. During that period, more than 62,000 individual farms, 8754 cooperatives, 578 joint-stock enterprises, and 1169 business partnerships emerged, occupying 94% of total lands (Baydildina et al. 2000). However, these changes did not affect the actual ownership and management practices, as in many cases the power remained in the hands of former state-farm managers (Petrick and Pomfret 2016). In the 1990s, because many farms faced bankruptcy, they were converted into limited liability partnerships (LLPs) during the pre-bankruptcy restructuring process. In these cases, former directors of production cooperatives often became directors and managers of these newly created entities. In addition, these directors were often not ready to operate farms under market conditions, as they lacked the necessary skills (Gray 2000).

Farm restructuring reversed the roles of large and small producers (OECD 2013). In 1997, large agricultural enterprises that were mostly former Soviet collective and state farms produced 43.8% of the total agricultural output, occupying 85% of all croplands. On the remaining 15% of the land, individual farms produced 10.2% of the total agricultural output, while individual households, with very small land plots, accounted for 46% of all agricultural output (Government of Kazakhstan 1998).

However, as a result of these reforms and the neglect of the agricultural sector due to the nation's economic decline following independence in 1991, capital investment in the agricultural sector almost ground to a halt, production significantly decreased, the rural population's living standards fell abruptly, and social tension in rural areas rose. During the first decade after independence, the sector incurred losses of 27.5 billion tenge (USD 435.1 million), while both crop and livestock production generated a net loss (−8 and −41%, respectively) (Tokbergenova et al. 2018).

Reform of Agricultural Finance and Support Systems

Although the *price liberalisation* was introduced in 1992, administered prices (set by the government and not through market forces) and state orders in the agricultural sector were abolished only two years later (OECD 2013). In 1995, the Food Contract Corporation (FCC) was established as a state grain purchasing agency, and state purchases continued to dominate the market. However, since the output prices remained low, many agricultural producers went bankrupt. A lack of working capital, obsolete machinery, and unavailability of microcredit to rural households (due to lack of collateral and poorly designed banking infrastructure) led to the government's decision to form a subsidised lending system for agricultural producers to enable them to continue operating.

In terms of *financial support*, between 1992 and 1994, the government provided funds to banks for low-interest lending to agricultural enterprises. From 1994, loans were extended through the state-owned Development Bank. However, the ongoing recession and growing agricultural debt forced the government to perform a significant financial restructuring of the agricultural enterprises between 1995 and 1997 (OECD 2013). The Fund for Financial Support to Agriculture (FFSA) was created in 1994 to transfer funds for supporting enterprises in the sector. By 1998, the development of lending structures and procedures became a priority of the nation's agricultural policy. KazAgroFinance (KAF) was established in the late 1990s to manage the machinery leasing programme, earlier performed by the FFSA. The Agrarian Credit Corporation was created soon after in order to develop a countrywide network of Credit Partnerships (OECD 2013).

Trade policy in the early 1990s mainly aimed at ensuring a sufficient food supply and mitigating food price inflation. Up to the mid-1990s, a rather non-restrictive import policy with zero tariffs and few non-tariff barriers for key agricultural goods was in place while it was complemented with tools aimed at restricting export, such as export quotas and licensing. In 1996, the situation somewhat reversed: import tariffs were established for a number of items, while export restrictions were relaxed. Since the mid-1990s, the government has been developing trade agreements with countries of the Commonwealth of Independent States[2] (CIS) (OECD 2013). Kazakhstan applied to join the World Trade Organization (WTO) in 1996. Although in the beginning, the accession negotiations progressed quickly, the process slowed down in 1998 after the introduction of ad hoc trade restrictions by the government (Feher and Fieldsend 2019).

At that time, the need to stabilise the difficult situation in the agricultural sector and strengthen the internal financial stability required the enhancement of state support to agricultural producers. The process of *financial rehabilitation and reorganisation* started in 1998 with the application of bankruptcy procedures for insolvent agricultural enterprises. In the period 1998–1999, the government introduced tax privileges with a 10% VAT rate on a number of agricultural products (instead of the general 30% VAT rate), and in 1998 a patent taxation system was created, which favoured agricultural producers by giving them an 80% tax reduction. The patent system of taxation is a special regime that allows farmers to pay simplified tax in advance, where the "patent" is the document issued after such taxes are paid (Tokbergenova et al. 2018). According to the OECD, in the period 1995–1997 Kazakhstan's Producer Support Estimate (PSE) was around 8% (meaning that agricultural support policies generated this share of the agricultural producers' gross revenue), which demonstrated the country's relatively moderate levels of support (OECD 2013).

[2] The Commonwealth of Independent States is an international organisation that includes some but not all republics of the Soviet Union. In the mid-1990s, the CIS included Azerbaijan, Armenia, Belarus, Kazakhstan, Kyrgyzstan, Moldova, Russia, Tajikistan, and Uzbekistan.

In the early 1990s, no mandatory *insurance* schemes existed in Kazakhstan, and only voluntary insurance for large-scale farmers was provided by private insurance companies. Compulsory agricultural insurance was then introduced by the Law on Insurance (Government Decree 1997). The government created KazAgroPolis as a state agency responsible for the provision of crop insurance. However, farmers did not show much interest in purchasing it, partly because the premium rates were rather high (20%). By 2000, the number of companies offering agricultural insurance had halved, and the initiative eventually failed (OECD 2013).

A *support programme* was started in the late 1990s to help farmers by decreasing the cost of inputs. A wide range of crops was covered by the scheme, including cereals, cotton, pulses, oilseeds, forages, some fruit, and vegetables. The subsidy covered 40% of seed production expenses. In addition, since 1997, support for the production of original, certified seeds was provided through the compensation of research and development (R&D) costs (OECD 2013). The pedigree livestock programme began at the same time, focusing on improving the quality of herds, and it incorporated various types of support (OECD 2013). These programmes allowed producers to diversify their agricultural portfolio, thus decreasing production risks. However, the overall impact on the diversification of the cropping pattern in the country was estimated as relatively low.

Water charges were introduced in Kazakhstan in 1994. The water usage fees were calculated based on volume and on the added value irrigation could bring, and varied across the country. However, the established prices were insufficient to fully cover the operational and maintenance costs of water companies that usually were state-owned. Although the introduction of water charges was a novel measure and low water charges were supportive of farmers, the income generated was insufficient and, without government support, water companies resorted to leaving the infrastructure to deteriorate. The first water user associations were established in 1996. Between 1996 and 2004, a state-financed project—the *Upgrading of Irrigation and Drainage Systems*—was implemented, covering 32,000 hectares of irrigated lands (World Bank 1996). The project was ultimately deemed a success, as it improved the irrigation and drainage systems, and introduced cooperative water use based on international best practice (OECD 2013).

Overall, during the latter years of the twentieth century, the decline in the country's agricultural production was reversed. However, farmers were still experiencing an economic decline, as well as a drop in rural population (due to migration to urban areas) and a level of social decline (due to the rise of social problems in rural communities), as they had to deal with the lack of available capital and financing options, unstable regulations, and limited access to the market (OECD 2013).

Turn of the Century: Agricultural Revival

During the first decade of independence, Kazakhstan went into recession, and this economic crisis deeply affected the agricultural sector. Vast agricultural lands were abandoned, as in a market economy it was no longer profitable for many farmers to cultivate them (Kraemer et al. 2015). The first years of transition were characterised by a drastic decrease in agricultural output (Fig. 3.1). The situation started improving only in the period 1999–2000, and since then the sector has been gradually recovering (Anderson et al. 2018). Crop yield fluctuations, changes in export prices, and such other factors have had a significant impact on the year-on-year agricultural output. However, more recently, physical volume of agricultural production has been showing slow growth. Agricultural gross production value (in US dollars, USD), on the other hand, experienced quite a rapid decline in 2013–2016. This clear fall (shown in Fig. 3.1) is largely explained by the devaluation of Kazakhstan's currency, the tenge (KZT), which happened in 2014, when the USD/KZT exchange rate jumped from around 150 to 180. In 2015, further devaluation of the

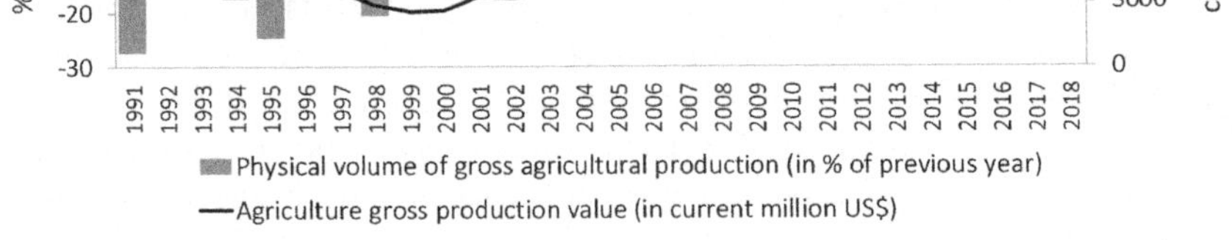

Fig. 3.1 Agricultural output: physical volume of gross agricultural production, 1991–2018 (*Source* Adapted by the authors from FAOSTAT [2020] and MONE [2020])

tenge occurred when Kazakhstan's National Bank decided to switch to a floating exchange rate between tenge and major world currencies. As a result, the USD/KZT exchange rate skyrocketed to 300, and later, in January 2016, reached its maximum (390 KZT for 1 USD) (Zholamanova et al. 2018). However, these fluctuations are not as dramatic when gross production value is calculated in tenge—in this case, the year 2014 was even characterised by a slight growth from KZT 2,386,103 million in 2013 to KZT 2,527,890 million in 2014 (Tokbergenova et al. 2018).

For a number of years after gaining independence until the beginning of the twenty-first century, the situation in the country was characterised by continuously declining agricultural output (Fig. 3.1 shows this as a percentage change compared to the previous year). The early 2000s showed some stabilisation, followed by a period of significant fluctuations (2007–2013) caused by a series of major droughts in 2010 and 2012 (Broka et al. 2016). However, since 2014, the sector reached some level of equilibrium with a small increase in agricultural production each year (physical volume) as percentage change compared to the previous year. Regarding agriculture production value in monetary terms, the sector experienced significant growth around the mid-2000s, followed by a decline that began around 2013–2014, which is explained by the devaluation of the national currency (Zholamanova et al. 2018).

Continuation of Land Reform

The 2001 Law on Land contributed insignificantly to land relations, as it failed to reinforce land tenure rights (determining who "holds" the land and who can use it, for how long, and under what conditions), or create favourable conditions for land market development (Law of the Republic of Kazakhstan 2001b). While agricultural land could be used only through a lease for up to 49 years, in 2003 the next stage of land reform began with the new Land Code that officially recognised private ownership of agricultural land (Land Code 2003). Nevertheless, farmers continued to use state-owned land, on a lease, probably because the rent prices were relatively low. In 2004–2005 land purchases spiked by 339% from 32,000 to 110,000 hectares, which accounted for less than 1% of the nation's total agricultural land (Gaisina 2011).

Over the years of land reforms, the organisational structure and distribution of agricultural land underwent significant transformation. In 1991,

99% of the total agricultural land was owned by corporate farms (almost 6000 such entities, including production cooperatives, companies, and partnerships). By 2007, they occupied only 51% of the land (just over 5200 entities, with an average 5000 hectares per farm), while individual farms took up 48% of the land (almost 170,000 entities, with an average land plot size of 200 hectares). The remaining 1% of the land was held by subsistence smallholdings (more than 2.2 million entities, with an average plot size of 0.2 hectares) (Gaisina 2011).

This artificial breakdown of large agricultural farms into smaller units had a two-sided effect. On the one hand, productivity of small individual farms was much higher due to a higher level of dedication from individual farmers to their family business, and as a result they accounted for two-thirds of the total national agricultural production volume. On the other hand, smaller farmers were less resilient to agricultural risks and shocks, and often they could not afford to adopt efficient, innovative technologies. This, first, led to unreliable returns, decreasing incomes, and has generally reduced the resilience of smaller farmers, thus hampering the social and economic sustainability of agriculture in Kazakhstan, and, secondly, impeded the sector's progress towards achieving the efficient use of inputs and ensuring environmental sustainability.

Development of the Agricultural Sector as Part of Economic Restructuring

In the early 2000s, the government began an active promotion of economic development (OECD 2013). In pursuit of economic modernisation, President Nazarbayev declared "economy first, then politics" as a guiding principle for the country's development (Euronews 2010). Such a position appeared to be somewhat opposite to the approach exercised during the first decade of the republic's independence when the focus was on consolidating political power. A highly centralised modernisation programme introduced by the presidential administration indicated significant interest in protecting agriculture and showed a sharp increase in spending for the sector (Petrick and Pomfret 2016).

The Agricultural Development Programme for 2000–2002 aimed at fulfilling the state priority of stabilising agricultural production by identifying and incentivising its competitive sub-sectors (growth points) (OECD 2013). In 2002, the Agriculture and Food Programme for 2003–2005 (AFP) was approved (Presidential Decree 2002), which marked the beginning of the active promotion of agricultural growth as part of the restructuring of the economy (Feher and Fieldsend 2019). Some of the AFP objectives focused on making national agricultural systems more efficient, improving market channels, and expanding state support for agriculture. Many subsidies and price support schemes were provided to make inputs (fertilisers, seeds, and fuel) more affordable, although some of the tools worked against resource-use efficiency (Pomfret 2007). Cheap inputs did not encourage farmers to save resources; often water and fertilisers, for instance, were applied in excessive amounts, causing soil erosion and water contamination.

Between 2000 and 2005, expenditure on agriculture was growing by 40% per year. The budget of the Ministry of Agriculture tripled between 2001 and 2005, and its share in the national budget increased from 2.5 to 6.5% (Feher and Fieldsend 2019). It continued to increase further, reaching KZT 139 billion in 2008 and KZT 200 billion in 2011 (compared to KZT 26 billion in 2001). The share of subsidies surged from 6% in 2001 to 24% in 2008, then to 39% in 2009 (Petrick and Pomfret 2016). Market price support—payments based on output and inputs used together—accounted for 82% of the total PSE in Kazakhstan in 2009–2011 (compared to 95% in the period 1995–1997). This decrease indicates that the support became slightly more oriented towards increasing the producers' efficiency and stimulating their productivity. However, an OECD report argues that such forms of producer support are predominantly production distorting and least efficient in increasing producer incomes, similar to the overall situation in emerging economies (OECD 2013). Market price support, for example, is directly linked to product output and can substantially affect production. Due to this fact, this type of support is considered to be production and trade distorting. Moreover, market price support is less effective in increasing producer income than support that is more decoupled from product output. Market price support also imposes additional cost on domestic consumers. At the same time, support delinked from product output, such as payments per hectare or direct income support, can be more effective in improving farmer incomes and achieving environmental objectives, while it has smaller spillover effect on international trade.

During the 2000s, the scope of agricultural policy instruments was substantially broadened. These policy instruments are discussed below.

Market Price Interventions

These were used mainly on the grain market and were performed by the FCC, to which farmers with a grain area of more than 250 hectares were obliged to sell their grain in quantities set annually by the government. Since the late 2000s, grain producers benefited from high purchase prices and export transportation subsidies. From 2008, a food price stabilisation policy was implemented aiming at mitigating price volatility on the global agricultural market. The policy included the creation of regional stabilisation funds. Meanwhile, local authorities were purchasing farmers' outputs and supplying these products to processors and retailers. Although this measure allowed the prevention of abrupt surges in food prices, it increased administrative intervention in agro-food pricing at the local level (OECD 2013).

Per Ton (Output) Payments

In 2006, these payments for livestock products were introduced for poultry, but by 2011 their coverage had gradually extended to virtually all livestock products, including beef, pork, lamb, milk, eggs, and wool. Payments were made per unit of output sold on the domestic market. The subsidy rates ranged from 13% (for eggs) to 50% (for beef) (OECD 2013). The total amount of payments for livestock output increased considerably between 2009 and 2011, reaching KZT 14 billion per year. No other regular support based on output was applied. However, in 2010, 30% of per-hectare payments for oilseeds and sugar beet were used to provide payments based on the output (OECD 2013). In order to be eligible for the payment, a farmer had to deliver the product for processing. This prompted farmers not only to plant the crops in order to receive high per-hectare subsidies, but also to harvest and market crops thereafter. However, this promising incentive was stopped after only one year of implementation.

Per-Hectare Payments

These were introduced in 2007 with the goal of diversifying crop production. They were applied to priority crops, including grains, oilseeds, sugar beet, forage crops, horticultural crops, cotton, and potatoes. They were calculated as estimated costs of producing these crops. Subsidy rates differed depending on the crop type, and increased rates were provided to producers who used moisture-saving technologies. The total amount of these payments varied over the duration of the programme and reached KZT 17.3 billion per year between 2008 and 2011 (OECD 2013).

Fertiliser and Chemical Subsidies

These subsidies were introduced in 2001. The subsidy supported the purchase of mineral fertilisers, chemicals, and herbicides. Initially, the cost of locally produced inputs was partially reimbursed to the manufacturers at a rate of 40% of the selling prices. In 2007, these subsidies were replaced by per-hectare payments. In both cases, the payment amount was delinked from the real input consumption and thus could not encourage efficient use. Moreover, in the late 2000s, additional ad hoc subsidies for fertilisers were provided together with per-hectare payments (OECD 2013).

Subsidised Lending

The use of this type of lending expanded, becoming available to more types of borrowers and financing more areas. Until 2003, lending to agriculture had doubled each year. However, in 2003, after the launch of AFP 2003–2005, it experienced a sharp, tenfold increase. Notably, the lending system of Kazakhstan was favouring large-scale agricultural producers. In 2007, these enterprises received 95.5% of the total loans allocated to the sector. This was despite the introduction, in 2005, of an "Agriculture Financial Support Fund" (a microcredit programme). The fund was led by the state-owned agency KazAgro and focused on providing support to low-income and small-scale farmers (those owning small plots of land on which subsistence crops and one or two cash crops are grown while relying almost exclusively on family labour). During the period 2005–2007, about 32,000 microcredits were allocated, meeting less than 2% of the demand for micro-financing (Gaisina 2011).

The volume of subsidised loans disbursed by KazAgro was rapidly increasing from the mid-2000s and accounted for KZT 59 billion per year in the period 2009–2011 (OECD 2013). Credit lines included both short-term (less than one year) and long-term loans (one to ten years). Interest rates depended on the type and purpose of the loan. For instance, in 2011, short-term loans for sowing and harvesting were provided at 4, 5, 6, 8, and 12% interest rates, whereas for the same commercial loan the interest rate was 12.3%. In 2011, long-term loans had interest rates from 4 to 9.5%, while commercial loans had interest rates from 10.5 to 11.5%.

Beginning in 2009, a significant amount of subsidised loans were allocated to state-funded investment projects, rather than supporting private entities (OECD 2013). This was owing to the government's concerns about the ability of private businesses to respond to the stated agriculture development goals in the face of increased instability in the global economy in the late 2000s. Of course, such a policy had implications for the development of private businesses in the country, especially of small and medium-sized enterprises (SMEs), which were being crowded out by the large state-owned enterprises.

Subsidised Leasing of Machinery
In 2011, the state-subsidised leasing fee for machinery had a 4% annual interest rate, which was a rather affordable rate even compared to non-subsidised leasing financed by KAF funds (9%) and by commercial companies (22–25%). This lower leasing rate made agricultural machinery (including innovative technologies) more affordable for farmers. By 2010, the share of agricultural leasing in the total leasing portfolio of the country was 92% (OECD 2013). Apart from leasing, the KAF also received KZT 1 billion both in 2003 and in 2004, and KZT 2 billion in 2005 to establish machinery service stations; since the disintegration of the Soviet Union this service had almost vanished. As a result, during these years (2003–2005), 25 machinery stations were established (OECD 2013).

Tax Benefits
Special tax regimes were introduced for agricultural producers with a new Tax Code, which was adopted in 2008 and entered into force in 2009 (Tax Code 2008). The social tax (i.e. the tax that an organisation must pay for the labour it uses) and the corporate income tax were cut from 30 to 20%. Legal agricultural entities (enterprises and cooperatives) benefited from a 70% reduction on six key business taxes (Feher and Fieldsend 2019). In 2008, the VAT rate on processed food was brought to 30% of the standard rate. As a result, compared to the standard VAT rate of 12%, agro-food processed products were eligible for a 3.6% VAT rate (OECD 2013).

Infrastructure Financing

Rural development and social issues started to receive attention from the government. The Law on State Regulation of Development of Agriculture and Rural Territories (Law of the Republic of Kazakhstan 2005) was adopted in 2005, as well as the Programme for Development of Rural Territories for 2004–2010 (Presidential Decree 2003). The funds of these programmes were directed towards improving the economic activity in rural areas and reconstructing the rural infrastructure (OECD 2013). From 2000 to 2002, KZT 2.2 billion per year was allocated to improving the water and irrigation infrastructure. From 2007, 205 state-funded investment projects were implemented, which focused on constructing and modernising agricultural facilities, with the grain, milk, poultry, beef, and horticulture sub-sectors accounting for 90% of the total investment (OECD 2013).

Research and Development

R&D was supported by the state agency KazAgroInnovation, which was created in 2007. It included 23 research institutes, functioning in all regions of the nation. KazAgroInnovation began developing rural advisory services for farmers in 2009 (OECD 2013). The budget for R&D substantially increased from KZT 770 million in 2001 to KZT 2.9 billion in 2009. However, it still accounted for only 0.3% of the agricultural value added (quite a low rate compared to the OECD countries, the majority of which stood at 1%, with a maximum of 4% in the USA) (OECD 2013).

General Services

Expenditures on general services for agriculture, especially phytosanitary and veterinary systems, increased from KZT 1.8 billion per year at the beginning of the 2000s to KZT 49.8 billion in the period 2009–2011 (OECD 2013).

New Institutions

In 2008, seven state institutions were consolidated under the aegis of KazAgro, including (i) the FCC and KazAgroProduct (KAP) were responsible for operating the grain/oilseeds and livestock markets, respectively; (ii) the Agricultural Credit Corporation (ACC) and the FFSA implemented lending at subsidised interest rates, and KAF implemented the state machinery leasing programmes; and (iii) KazAgroMarketing

(KAM) provided price setting and marketing services, and KazAgroGarant (KAG) was a guarantee fund for warehouse receipts[3] (FAO 2017). Besides implementing support programmes, these agencies also performed commercial operations and were the dominant players on the market (OECD 2013).

Trade Policy Instruments and Agreements

Import Tariffs
Several types of tariffs were implemented: ad valorem (proportionate to the estimated value of the goods concerned) with 0% rate for non-carded cotton and raw cane sugar, 5% for wheat, 15% for eggs, and 20% for powdered milk. White sugar had a specific tariff of USD 340 per ton. Meat imports were under the tariff rate quotas regime (where certain quotas in the form of specific amounts were set for different types of products, for which the tariff was zero, while for the amounts above the quota the tariff was up to 75%).

Temporary Export Bans
Temporary export bans were used in case of a need to stabilise the prices, mainly for oilseeds.

Export Subsidies
The transportation of grain through the territories of Russia and China was subsidised—in 2009, at the rate of USD 20 per ton; in 2011, at USD 40; in 2012, at USD 27—until it was finally abolished in 2012. A total of KZT 26.8 billion was spent on these subsidies between 2009 and 2012 (FAO 2017). In 2008, Kaznex Invest was established to support agricultural producers who were exporting their goods. During the period 2010–2013, KZT 494 million were spent on reimbursing their export costs. In 2019, KazExportGarant was created to protect exporters against financial risks (FAO 2017). As a result of these export-promoting measures, the annual volume of wheat exports almost doubled and its

[3]Warehouse receipts are an innovative financial instrument that allows to minimise risks for farmers. A commodity (harvested crop yield, for instance) is being delivered by the farmer to the specialised warehouse against the receipt that can serve as collateral for a bank loan (Law of the Republic of Kazakhstan 2001a).

annual value more than tripled during the first decade of the twenty-first century (OECD 2013).

Regional Trade Agreements

During this period, Kazakhstan took certain steps towards Eurasian integration. In 2007, an agreement for the creation of a Customs Union with Belarus and Russia was signed, which came into effect in 2010 (FAO 2017). For Kazakhstan, this triggered an increase in tariffs for agricultural produce and the introduction of quotas on meat, detailed above (OECD 2013). After recovering from a slowdown in 1998, the negotiations on the WTO accession lost momentum again in the late 2000s. The introduction by Kazakhstan, in 2008, of export restrictions and agricultural subsidies created more potential obstacles to accession (Feher and Fieldsend 2019). Due to the creation of the Customs Union, imports of agricultural products from Belarus and Russia increased. At the same time, Kazakhstan's exports shifted towards the Central Asian region, Azerbaijan, Turkey, Iran, and North Africa. Imports have been increasing more rapidly than exports, due to improvements in incomes and the strengthening of the national currency.

Agricultural Risk Management Policies

The *Law on Mandatory Crop Insurance*, adopted in 2004 (and amended in 2011), enforced mandatory crop insurance for grains, oilseeds, sugar beet, and cotton (Law of the Republic of Kazakhstan 2004). Insurance was provided by private insurance companies and covered crop damage from such events as drought, heavy rains, floods, hail, frost, torrents, and tornadoes (OECD 2013). The insurance premium was fixed by law and the insurance payments were calculated based on producer losses (Feher and Fieldsend 2019). The state subsidised 50% of the payments made by the 42 insurance companies participating in the scheme. In 2011, the Fund for Financial Support to Agriculture spent KZT 1.1 billion on this programme (OECD 2013). The number of insurance policies rose until 2009 but dropped thereafter, possibly because of difficulties producers faced when settling indemnities with the insurance companies (Feher and Fieldsend 2019).

Warehouse receipts were introduced in Kazakhstan in 2001 under the European Bank for Reconstruction and Development (EBRD) project and were legalised in the Law on Grain, which obliged grain-receiving enterprises to operate in accordance with the guarantees provided by the grain warehouse receipts (Law of the Republic of Kazakhstan 2001a). This financial instrument was aimed at being used as collateral to obtain loans from commercial banks with interest rates of 10–11% in the period 2003–2006, compared to the standard rate of 12.5–14.5% (OECD 2013). The total amount of such loans increased from KZT 2.4 billion in 2003 to KZT 225.9 billion in 2011 and accounted for over 30% of the total agricultural loans in 2010. In the same year, out of the 222 licensed grain storage facilities, 51 participated in the grain receipts scheme, operated by KAG, which acted as a guarantor (OECD 2013). In 2008, grain warehouse receipts for cotton were introduced. However, these were not successful due to the fact that cotton producers do not store cotton but prefer to receive cash for their product on the spot (OECD 2013).

Attempts to Mainstream Sustainability and Environmental Aspects in Agricultural Policy

Since 2005, the Agricultural Law covered environmental and social issues and included the objective of ensuring sustainable development of agriculture and rural areas (Law of the Republic of Kazakhstan 2005). Table 3.1 shows the variety of agricultural programmes and policy documents that included environmental and sustainability priorities. Areas that received most attention are water management, rangeland (land used for grazing animals but with wild vegetation), and pastures (land sowed with grass and other short plants suitable for grazing animals, especially sheep and cattle). However, integrated pest-management methods were not really considered (OECD 2013).

Prior to 2007, many policy instruments adversely affected farmers' input use efficiency; in particular, limits on prices for diesel fuel (10% discount in 2010–2012), subsidies for water delivery (40%), and reduced cost of fertilisers, chemicals, and fuel (OECD 2013). After 2007, due to growing concern over soil degradation and water overuse related to existing crop growing practices, many instruments were substituted with more sustainable ones, creating incentives for more efficient use of available resources and the adoption of sustainable agricultural practices. Subsidie s for the use of fertilisers, chemicals, and fuel were replaced by per-hectare payments, which also were meant to enhance crop diversification with oilseeds and horticultural crops (OECD 2013).

Table 3.1 Environmental and sustainability priorities in Kazakhstan's agricultural policy between 2000 and 2011

Policy documents	*Priorities*
State Agricultural Food Programme for 2003–2005	• Soil conservation and rehabilitation of soil fertility • Improvement of laws that encourage sustainable soil use
Programme for Rational Use of Agricultural Land for 2005–2007	• Measures to protect and rationally use land resources • Soil conservation and restoration of soil fertility
The Concept of Sustainable Development of Agricultural Sector for 2006–2010	• Increased use of organic fertiliser to improve soil fertility • Enhancement of crop rotation • Expansion of perennial legume grasses
The Strategy for Development of Agriculture until 2010	• Improvement of soil fertility and reduction of soil degradation and desertification through the introduction of innovative production technologies
Strategic Plan for Development of the Republic of Kazakhstan until 2010	• Use of sustainable, organic and environmentally friendly agricultural practices
The Applied Research Programme for the Agro-industrial Complex for 2009–2011	• Creation and improvement of stress-resistant crop varieties and resource-saving technologies • Development of machinery complexes for moisture- and energy-saving technologies and systems of water supply for remote farms • Rational management of natural resources

Source Adapted by the authors from OECD (2013)

The water delivery discounts varied between 25 and 90% depending on the type of irrigation system used, with the highest rate applied to users of drip irrigation and reduced tillage. As a result, moisture-saving techniques (such as saving soil moisture by crop covering and reduced tillage) were applied on 11.7 million hectares in 2011 (72% of the area under grains), compared to only 1.1 million hectares (8%) in 2003. The Ministry of

Agriculture has also set a goal to increase the area under reduced tillage[4] to 80% by 2014 (OECD 2013). The data showing whether this goal has been achieved are unavailable, or these data are not in the public domain, which suggests that most likely these estimates have not been calculated, possibly owing to difficulties with measuring and/or unavailability of the required technology.

Current Times: Is Kazakhstan Heading Towards Sustainability?

In 2012, President Nazarbayev announced the *Kazakhstan-2050* strategy (Message from the President 2012), which included numerous objectives aiming at making the nation one of the 30 most developed countries in the world by 2050. The strategy addressed issues such as sustainable development and a new policy on the management of water resources in the context of the rising needs of the agricultural sector (Baubekova and Kvasha 2019). The main objectives of the strategy relating to agriculture are the modernisation of the agricultural sector, the provision of support to small agricultural producers, and the strengthening of the water resources policy (Message from the President 2012). Since the strategy was introduced, a number of development programmes in the agricultural sector have been adopted to fulfil these goals. These policies were formulated using a distinct top-down approach, with a high degree of centralisation in setting the policy objectives (OECD 2013).

Continued Focus on Agricultural Development

In order to implement the overarching *Kazakhstan-2050* strategy, a series of strategic plans (per decade) and development programmes (per five-year period) have been introduced. In one of them, the *Accelerated Industrial and Innovative Development of the Republic of Kazakhstan for 2010–2014* (PEIID), the agricultural sector was one of the priorities for diversification (Presidential Decree 2010). Sectoral programmes were developed to implement the PEIID, including the *Programme for*

[4] Reduced tillage is a practice of minimising soil disturbance and allowing crop residue or stubble to remain on the ground instead of being thrown away or incorporated into the soil. It reduces the amount of inputs, such as labour, energy, and water (due to preserving soil moisture), as well as helping to prevent soil erosion (OECD 2013).

the Development of the Agro-Industrial Complex for 2010–2014 (Government Decree 2010) with food security as the main objective and with the specific target of achieving 80% self-sufficiency in basic food items (OECD 2013).

In 2013, the subsequent *Programme for the Development of Agro-Industrial Complex in the Republic of Kazakhstan in 2013–2020* was adopted. The programme's primary objective is to create an enabling environment for increasing competitiveness of agricultural producers. It includes the provision of loans with a subsidised interest rate and guarantees, financial support to cooperatives, and human resources development (Feher and Fieldsend 2019). Between 2013 and 2020, spending within this programme is expected to be KZT 3.1 trillion (FAO 2017). Over 40% of the programme's budget is used to finance output and input subsidies and per-hectare payments (OECD 2013). However, resource-use efficiency incentives or other sustainability-related measures are not included in this programme (Feher and Fieldsend 2019).

In addition, another programme—the *State Programme for Agro-Industrial Development for 2017–2021*—was launched. It aims to double the gross output of the agricultural sector, enhance the grain sector profitability by 30–40% by implementing standards for organic food products, introduce grant distribution schemes, and encourage crop diversification from wheat to maize and soybeans (MOA 2019).

The *Development Strategy of KazAgro for 2011–2020* sets the goal of increasing agricultural financing and support, ensuring crop diversification, and expanding farmers' access to loans (Government Decree 2011). In 2016, KazAgro spent KZT 72.6 billion for these purposes (FAO 2017).

In 2011, the *Programme on the Development of Beef Export Potential for Kazakhstan* was launched, and in 2013, the *Master Plan for Beef Cattle Sector Development until 2020* was approved, providing support for imports of foreign cattle and the improvement of the genetic stock of livestock. Also, since 2013, livestock producers have been provided with cost reimbursements and cost subsidies for high-quality feed (FAO 2017).

In 2013, a *Regional Specialisation Scheme* for Kazakhstan was prepared, proposing the types of agricultural production most suitable for each region of the country, based on the local context. The incentives that follow recommendation include support payments and access to subsidised loans. This programme was planned to be implemented in stages. By 2020, beneficiaries who fully comply with the scheme's requirements should be eligible for different types of assistance (Petrick and Pomfret 2016).

In 2015, the government amended the land legislation, facilitating access to land for foreign investors by increasing the lease term to 25 years (compared to 10 years stipulated in the Land Code of 2011) and relaxing tax administration. However, after public protests, a moratorium on the amendment was introduced from 2016 to 2021. Additionally, sales of state land were suspended for the same period. As a result, progress towards competitive land markets has stalled (Anderson et al. 2018). Furthermore, the amendment also terminated the lease of agricultural land by Kazakh residents (including foreigners living in the country), substituting it with the auction-like mechanism for purchasing land plots (FAO 2017).

Currently, an *Irrigation and Drainage Improvement Project*, partially financed by the World Bank (2015–2022), with a total USD 343 million in funding, is under implementation in southern Kazakhstan in order to improve irrigation and drainage services (Petrick and Pomfret 2016). Investment projects that are part of *Development of Regions and Business Roadmap 2020* programmes receive state support in the form of interest rate subsidies on loans for investment (KZT 1.6 billion between 2010 and 2012) (OECD 2013).

The gross output of agricultural production (monetary value of crop production plus that of animal husbandry) in Kazakhstan continued to increase in the period 2010–2016 (Fig. 3.2); its average annual value is about KZT 2.2 trillion. The livestock sector has been demonstrating only positive dynamics of production growth with an average increase of 10.4% per year. From 2010 to 2015, crop production and livestock accounted for 52.5 and 47.2%, respectively, of all agricultural production (Tokbergenova et al. 2018).

Both types of production exhibit an upward trend through the second decade of the twenty-first century. While animal husbandry experiences a steady increase, crop production had faced more fluctuations between 2010 and 2013 and a more flattened trend up to 2015. The years 2010 and 2012 were both characterised by quite poor crop yields (2010 was the worst in the last decade with 12.1 million tons of grain, including rice, and legumes cultures collected), while 2011 was the year of the most abundant harvest over the last years (26.9 million tons) (MONE 2020).

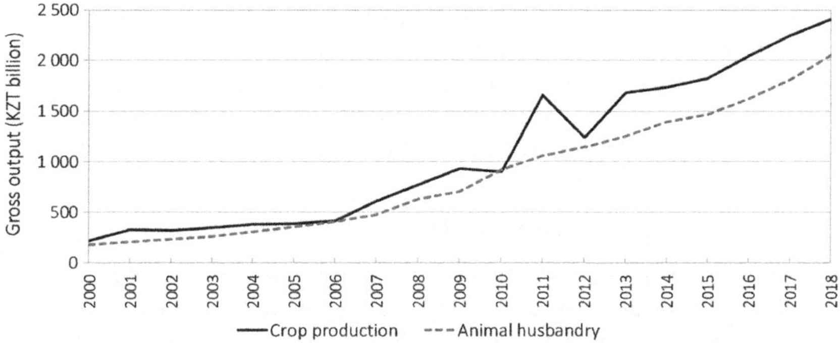

Fig. 3.2 Annual growth of crop production and animal husbandry in Kazakhstan, 2000–2018 (*Source* Adapted by the authors from MONE [2020])

The main novelties in policy instruments during the second decade of the twenty-first century can be summarised as follows.

Changes in Tax Regimes

Until 2015, farmers benefited from a 70% reduction on six taxes: land tax, property tax, social tax, VAT, enterprise income tax, and tax on vehicles. After 2015, due to the adoption of amendments to the Tax Code, a 70% discount on land tax was discontinued, and land tax rates increased five times (Law of the Republic of Kazakhstan 2014). Currently, instead of making payments for all six taxes, individual farms (now limited to 3500 hectares) have to pay a Single Land Tax, which is 50% higher than the six separate taxes combined. Further, if the annual income of an individual farms exceeds KZT 150 million, then the standard 10% tax rate for agricultural producers applies (Petrick and Pomfret 2016).

Changes in the Lending System

Starting from 2013, budget funds that were previously exclusively transferred to KazAgro's lending agencies were partially re-allocated to the other lending institutions for the provision of subsidised agricultural loans. This increased the amount of subsidised loans made available to farmers. Three new subsidised lending programmes, initiated in 2014, provided loans for smallholder farmers for sheep purchase, horse breeding, and water network development for pastures (Petrick and Pomfret 2016).

Overall, these policies resulted in steady increases in crop and livestock production, a significant diversification of farmers' crop portfolio, and an enhanced role for private lending institutions in the country. Furthermore, smallholders, on one hand, enjoyed improved access to financial resources, but on the other, suffered from an increased tax burden.

Agricultural Debt Restructuring

The crises of the late 2000s (first the local banking crisis in 2008 and then the global financial crisis in 2009) have aggravated the farm debt situation. The ban on grain exports in 2008 was a reactive measure against the global food price surge. It caused significant losses to the grain sub-sector, which in 2009 also faced low prices, following the abundant harvests in countries of the Commonwealth of Independent States, and suffered from a drought in 2010. The share of bad loans increased sharply in 2010, and by 2012 accounted for over half of the total disbursed agricultural credit in both commercial banks and KazAgro financial agencies (OECD 2013).

This led to the government's decision to release an agricultural debt relief package of USD 2 billion. The agricultural programme for 2013–2020 introduced a range of measures to address the debt situation in the sector, providing an extension of up to nine years for overdue loans and offering a subsidy to the interest rate for these loans, bringing the effective interest rate to 7% (a significant decrease from an average 14%). In 2016, KZT 2.9 billion was spent on writing off fines and penalties on overdue loans (Petrick and Pomfret 2016).

In order to encourage commercial banks to participate in long-term lending to agriculture, KazAgro should provide them with funds to be used for lending to farmers. This will likely result in reducing the reliance of agricultural producers on state agencies for subsidised loans and developing a more open financial market. More flexibility could also be achieved through the introduction of a new policy of interest rate subsidies instead of the application of a rigid fixed rate (OECD 2013).

The Eurasian Economic Union and the WTO Accession

In the second decade of the twenty-first century, intending to further promote agricultural exports and trade, the government took steps to participate in the Eurasian Economic Union (EEU), and to accede to the WTO (FAO 2017). The EEU became a successor of the Customs Union of Belarus, Kazakhstan, and Russia in 2015. The new union created a common market space for Kazakhstan, Russia, Belarus, Armenia, and the Kyrgyz Republic; intensified competition; and provided unified technical regulations (including 16 regulations relating to the agro-food sector); veterinary and phytosanitary requirements and other measures (Petrick and Pomfret 2016). In 2014 and 2015, no restrictions—such as minimum export prices, import bans or quotas—were used for agricultural goods within the EEU. The limit for an import tariff was set at 30% (with the average set at about 17%) (FAO 2017).

After 20 years of negotiations, Kazakhstan became a member of the WTO in 2015 (Petrick and Pomfret 2016). The agricultural sector was one of the impediments in the negotiations, particularly the country's market-regulating policies, state support for agriculture, and the absence of proper food safety measures and marketing standards (Feher and Fieldsend 2019). The new commitments remove the possibility of imposing duties on agricultural export (except on skins and wool). At the same time, import tariff rates for all agricultural products are to be decreased from 16.7 to 7.6% on average. The state support subsidies should not exceed 8.5% (compared to the 5% to which other WTO members have to adhere) (FAO 2017).

Sustainability and Climate Change Adaptation Trends in Kazakhstan's Agricultural Policy

In 2012, President Nazarbayev in the *Kazakhstan-2050* strategy emphasised "food insecurity" and the "depletion of natural resources" among the key global challenges of the twenty-first century (Message from the President 2012). In the same strategy, special attention was paid to innovations in the agricultural sector and supporting smallholder farmers. Since then, several incentives have been introduced for farmers to improve resource-use efficiency and increase consumption of local food (FAO 2017). However, agriculture in Kazakhstan is highly vulnerable to climate change, namely frequent droughts and water shortages that affect crop productivity (FAO 2017). One of the country's approaches to tackle climate change has been to focus on the development of low-carbon and renewable energy sources.

A number of policies have been adopted to lead national climate change mitigation and adaptation actions. The *Concept of Transition of the Republic of Kazakhstan to a Green Economy 2013–2020* focuses on the rational water use and the proliferation of renewable energy technologies (OPRK 2013). At the same time, the *National Programme of Accelerated Industrial and Innovative Development for 2010–2014* emphasises integrated water management and land conservation by adopting advanced irrigation methods (FAO 2017).

Another way in which agriculture is adapting to climate change is related to increased crop diversification, and several measures have already been implemented in this regard. The *Crop Diversification Policy* was introduced in order to reduce the area used for wheat production, and increase the area used for forage crops and oilseeds (FAO 2017). It also envisaged the increase of the share of oilseed crops that are drought-tolerant and that provide a variety of high-value products, increasing farmers' incomes. The larger share of oilseed crops is needed in order to expand the variety of crops rotated in a land area, pursuing the goal of maintaining soil quality and reduce the risk farmers face due to other crops failing. The policy was implemented through the wide use of direct subsidies for cultivating these crops, which resulted in an increase of total oilseed area from 0.67 million hectares in 2007 to 2.04 million hectares in 2016 (FAO 2017). However, the success of this subsidy scheme was considered insufficient for stimulating crop diversification, so in the new *Agricultural Development Programme for 2017–2021* these subsidies were restructured, keeping the subsidies for crop-processing of oilseeds, and terminating per-hectare subsidies for all other crops (wheat, barley, potatoes, vegetables, etc.) (FAO 2017).

The *"Healthy Lifestyle" Programme* was adopted for 2008–2016 and included increased consumption of fruits and vegetables among its main objectives, which not only had a positive effect on people's health but also most likely increased production of these categories of crops due to higher demand. Thus, the programme contributed to crop diversification (FAO 2017).

The Way Forward: Addressing Emerging Issues

Despite the significant progress made towards the revival of the agricultural sector in Kazakhstan since the nation gained its independence, many challenges that hamper its sustainable development and growth still exist. These barriers, as well as issues that might emerge in the future—for instance, those connected to climate change—should be carefully examined and addressed. The following recommendations could help

Kazakhstan achieve its agricultural growth objectives more efficiently and effectively.

Improve Policy Formulation, Transparency, Enforcement, and Implementation

Kazakhstan's centralised approach to policymaking in general, and in relation to the agricultural sector in particular, has still much in common with traditional Soviet models (Petrick and Pomfret 2016). Some policy goals are poorly articulated and their implementation is requiring significant bureaucracy (Feher and Fieldsend 2019). Furthermore, certain aspects, such as protecting the environment, require stronger enforcement. Information on the availability of and access to different support schemes is poorly communicated to the farmers. Even when subsidies or other support mechanisms are available, their implementation is often delayed (e.g. arriving after the farmer has purchased inputs for sowing and fertilising). In addition, transparent rules on subsidy schemes often appear to be discretionary when applied at the local level (Petrick and Pomfret 2016).

There is also a need to create an enabling environment for the successful development of private initiatives in agriculture by introducing measures beyond cheap access to inputs and capital. New measures are needed to entice private sector actors to become involved in profitable opportunities among sustainable options available. Therefore, it is imperative to refocus policy efforts on strategic investments, with the active engagement of private firms, to support sustainable agricultural development.

Shift the Policymaking Approach to Ensure Stakeholder Involvement in the Decision-Making Process

The recently created Business Council acts as a platform for interaction between the government and private stakeholders. Nonetheless, although stakeholder engagement is high on the agenda of all governmental bodies, the top-down approach in policymaking in Kazakhstan leaves little room for different groups of stakeholders to influence the decision-making process. Thus, there is a need to ensure that all stakeholders are represented and involved at all stages of the policy cycle (OECD 2013).

Improve Land Tenure

After years of continuous land reforms, land markets remain weak, which constrains the use of land as collateral. Although the Land Code of 2003 established the private ownership of farmland, the land purchase process has not made much progress so far, due to the high prices of state land (Gaisina 2011). Efforts should focus on enforcing the implementation of land reform objectives (i.e. to achieve full and reliable privatisation of agricultural land in order to increase agricultural productivity) and on overcoming regulatory and administrative constraints in the purchasing of agricultural land.

Strengthen Institutions

Some institutional settings in Kazakhstan are inadequate for the rational development of the agricultural sector. Weak market institutions do not permit the absorption of price shocks, rendering the improvement of the functioning of these institutions crucial. A number of state institutions receive only public financing and have accumulated significant power over the sector. For instance, the activity of the Food Contract Corporation on the grain market is excluding private traders due to its market intervention and prior claims on elevator space and transportation (OECD 2013). Furthermore, it does little to help farmers increase their profits by improving supply chains or to create international awareness of Kazakhstan's quality standards (Feher and Fieldsend 2019). Subsidised interest rates and leasing fees provided by KazAgro's credit agencies are also available through commercial banks, but by prioritising specific producers the process marginalises independent commercial loans for projects with high-return potential (Petrick and Pomfret 2016) while having little concern for allocative efficiency (Feher and Fieldsend 2019). There is a need to strengthen and/or re-orient existing state agencies and introduce more market-based institutional arrangements to support the development of private institutions.

Improve Financing and Prevent Indebtedness

The level of bank lending in the agricultural sector of Kazakhstan remains low (Petrick and Pomfret 2016). The key problems are the lack of collateral and strong government intervention in the rural financial market.

In addition, there is a risk that subsidised loans will be overused by small-scale producers for purposes other than the development of agricultural activities or the purchase of new technologies. This might happen due to the lack of control over the actual use of loans, thus creating indebtedness among farmers. It is therefore important to differentiate subsidies based on types of activities and to enhance control over their actual use. Further, as noted earlier, there is a need to create enabling conditions for private investment in agriculture. Finally, it is necessary to develop special borrowing opportunities for agricultural producers, who implement sustainable and climate-smart agricultural practices, beyond the currently used water-saving technologies.

Continue Improvements in Physical Infrastructure

While more resources have been directed to infrastructure development in recent years, such as investments in the establishment of veterinary laboratories and the introduction of modern quality control systems in the processing sector (Saparov et al. 2018), much more has to be done. Modern phytosanitary, veterinary, and food safety systems still need to be fully developed. Furthermore, there are deficiencies in critical transport and irrigation infrastructure. Larger investments should therefore be made in these areas (OECD 2013).

Support Smallholder Farmers

Efforts to develop large-scale agriculture should be complemented by helping small farms integrate into local supply chains. This should be pursued through the promotion of vertical arrangements between smallholders and enterprises that process agricultural products, and through further development of lending opportunities and other support mechanisms directed at small agricultural producers.

Develop an Efficient System of Rural Advisory Services

A system of advisory services is just starting to develop in Kazakhstan. Coupled with well-coordinated education and R&D, these services can contribute to strong productivity growth and widespread adoption of sustainable farming practices. Several key commodity sectors in Kazakhstan, including vegetables, potatoes, meat, milk, wool, and cotton, are currently dominated by households and individual farms (OECD 2019). They typically lack sufficient technology and the necessary business experience. Targeted rural advisory services would help them integrate into agricultural markets.

Improve Risk Management Options for Farmers

Kazakhstan's producers face significant risks related to climate, animal health, pests, and diseases, yet risk management in agriculture is not an apparent focus for the nation's government policy (OECD 2013). There is a need to develop a comprehensive set of risk management options (e.g. index-based crop insurance) for agricultural producers, especially for smallholders, and to assess the current crop insurance scheme to improve its performance.

Create Incentives for Sustainable Practice in Agriculture

Low returns in agriculture and insufficient credit lines for small farmers leave them unable to invest in sustainable agricultural technologies and to bear the risk of adopting innovations. Such measures as current subsidies on fertilisers, chemicals, and irrigation are adversely affecting the environment and should be abolished. Instead, disincentives should be introduced for harmful agricultural practices. Furthermore, land taxation could be improved by further differentiating between farmers applying conventional production methods and those using environmentally friendly techniques.

Conclusion

Agriculture plays an important role in ensuring Kazakhstan's food and economic security, as well as supporting employment and settlement potential in rural areas. Although the share of the agricultural sector in the nation's GDP has substantially decreased since the Soviet era, agriculture remains an important part of the country's economy, employing about a third of its workforce. In the *Kazakhstan-2050* strategy, the government identified the agro-industrial sector as key to the nation's economic transformation in order to ensure the country's sustainable development and equitable growth (Message from the President 2012).

Since Kazakhstan gained its independence, agriculture has undergone considerable change in its overall performance and numerous policies have been implemented. During the 1990s, the sector experienced serious economic shocks, coupled with substantially decreased state support and incomplete land reform. Agricultural policy and programmes were largely ineffective, and the sector experienced a deep recession.

After the turn of the century, the revival of the sector began. Numerous agricultural development programmes were introduced, which included enhancing the efficiency and competitiveness of the agro-industrial complex. As substantial public funds were directed to the sector making land tenure arrangements (i.e. legal assurances of land ownership and land use) more transparent, agricultural performance improved significantly.

Beginning in the late 2000s, the scope of agricultural policy objectives has broadened to cover environmental and social dimensions. In the second decade of the twenty-first century, sustainable development as the objective of the agricultural sector gained momentum and the need for agriculture to adapt to changing climatic conditions was recognised.

Since 2014, the international perspective has gained more importance, triggering the formulation of new objectives, such as the increase of livestock exports and the adaptation of Kazakhstan's agricultural sector to open market conditions in view of the nation's membership in the EEU and its WTO accession in 2015. Furthermore, the launch of the One Belt and One Road Initiative of the People's Republic of China provides a great opportunity for transit trade through Kazakhstan and increased exports by non-oil sectors including agriculture.

With its vast areas suitable for agriculture, Kazakhstan has great potential to be a regional leader in food exports. However, difficult climatic conditions and historical structural challenges are hampering the full-scale development of the agricultural sector. Table 3.2 shows the key obstacles and improvement opportunities that should be considered in order to ensure the sector's sustainable development.

The country is facing the challenge of becoming internationally competitive on the global agriculture market without turning to environmentally harmful and unsustainable practices. Large areas of formerly fertile land are now unsuitable for farming due to historical contamination and pollution. Such issues have already been addressed in the long-term programmes that aim to develop the country's agro-industrial complex until 2027. For example, the main long-term priority is the meat-producing livestock based on the tradition of nomadic remote

Table 3.2 Summary of recommendations and directions for improvement

Direction for improvement	*Its significance and aim*	*Benefits*	*Potential barriers*
Coping with current and future risks, including those posed by climate change	To cope with production risks posed by the changing climate (droughts, floods, water availability) and price risks caused by the high volatility of the international market and sudden exchange rate fluctuations	Improved resilience of agricultural producers to future changes and shocks, including climatic	Limited availability and access to technologies and financial resources; little interest of insurance companies to provide crop insurance
Ensuring introduction and upscaling of climate-smart technologies	To make use of targeted subsidies to ensure introduction and wide-scale adoption of climate-smart technologies; launch initiatives to further improve wheat and livestock productivity, including increasing support to research institutions to enhance the selection of improved crop varieties and animal breeds and to promote agricultural diversification	Improved agricultural sustainability in three ways: economic (higher productivity and returns from farming), social (increased farmer's resilience) and environmental (improved input and resource-use efficiency, less natural degradation)	Limited availability of and access to financial resources; difficulties with knowledge dissemination; limited availability of climate-smart technology
Increased availability of loans for farmers	To broaden the scope and increase the volume of concessional loans for agricultural producers, especially for smallholders and those who use innovative resource-efficient technologies; to better inform farmers about the existing opportunities and procedures for receiving loans	Improved access to financial resources for investment in agriculture. Improved affordability of innovative technologies	Even a relatively low interest rate of concessional (subsidised) loans may inhibit the willingness of farmers to receive a loan because of the high risk involved and of the low capacity of smallholder farmers to bear this risk

Direction for improvement	*Its significance and aim*	*Benefits*	*Potential barriers*
Attracting private investment in sustainable agriculture	To identify effective mechanisms of attracting private investment in sustainable agriculture and explore whether public–private partnership schemes are suitable for the sector	Increased financing of the sector, broader implementation of innovative technologies, higher productivity	Investors show limited interest in the sector because of comparatively low returns from agricultural activities
Improving existing risk management mechanisms	To increase accessibility to seasonal finance and improve agricultural insurance programmes	Farmers' risk related to agricultural production is reduced	Possible difficulties with broadening crop insurance are linked to low interest of insurance companies in providing such insurance because of high weather and climate uncertainties
Improving education and knowledge management	To create awareness among stakeholders about the climate change impact on agriculture; promote best available climate-smart practices and technologies; improve rural advisory services; introduce e-agriculture	Knowledge required to follow sustainable agriculture path is updated and widely disseminate	Low level of R&D funding and absence of public rural advisory services might impede achieving this goal

(continued)

Table 3.2 (continued)

Direction for improvement	*Its significance and aim*	*Benefits*	*Potential barriers*
Supporting small individual and family farms	To develop additional opportunities for receiving concessional financing for smallholders and improve their accessibility; to form a new business model for the farms of less than one hectare, which would permit them to work as cooperating units	Improved financial resources of smallholder farmers Enabling conditions are created for farmers to share agricultural costs and risks, and to benefit from enhanced affordability of innovative technologies and machinery	Current policy trends of favouring large-scale producers and high level of competition from their side are likely to be a barrier. Farmers might be reluctant to unite in any type of cooperative-like units based on past negative experience

Source Compiled by the authors

pastures (MOA 2019). However, although socio-economic and environmental concerns are mentioned in recent agricultural policy documents, they do not appear to be a high priority when it comes to practical implementation. To address these issues, further amendments in the policy implementation of the reforms laid in the *2025 Strategic Development Plan* (adopted in 2018) are required (Government of Kazakhstan 2018). It is imperative to overcome these barriers to achieve the sustainable development of Kazakhstan's agricultural sector in the future.

References

Anderson, K., Capannelli, G., Ginting, E., & Taniguchi, K. (Eds.). (2018). *Kazakhstan: Accelerating economic diversification*. Mandaluyong: Asian Development Bank.

Baubekova, A., & Kvasha, A. (2019). Implementing water-related sustainable development goals. In S. Xenarios, D. Schmidt-Vogt, M. Qadir, B. Janusz-Pawletta, & I. Abdullaev (Eds.), *The Aral Sea Basin: Water for sustainable development in Central Asia* (pp. 197–222). Abingdon: Routledge.

Baydildina, A., Akshinbay, A., Bayetova, M., Mkrytichyan, L., Haliepesova, A., & Ataev, D. (2000). Agricultural policy reforms and food security in Kazakhstan and Turkmenistan. *Food Policy, 25*, 733–774.

Broka, S., Giertz, A., Christensen, G., Rasmussen, D., Morgounov, A., Fileccia, T., & Rubaiza, R. (2016). *Kazakhstan agricultural sector risk assessment* (World Bank Group Report Number 103076-KZ). Washington, DC.

Deshpande, R. (2006). Land reform and farm restructuring in Central Asia: Progress and challenges ahead. *Policy reforms and agriculture development in Central Asia* (pp. 131–170). Boston, MA: Springer.

Euronews. (2010). Interview "Nazarbayev: 'Economy first, then politics'". Available at https://www.euronews.com/2010/01/15/nazarbayev-economy-first-then-politics. Accessed 12 December 2019.

FAOSTAT. (2020). *Value of agricultural production*. Food and Agriculture Organisation of the United Nations. Available at http://www.fao.org/faostat/en/#data/QV. Accessed 30 January 2020.

Food and Agriculture Organization (FAO). (2017). *Country fact sheet on food and agriculture policy trends: Kazakhstan*. Food and Agriculture Organisation of the United Nations. Available at http://www.fao.org/3/a-i7676e.pdf. Accessed 17 November 2019.

Feher, I., & Fieldsend, A. F. (2019). *The potential for expanding wheat production in Kazakhstan: Analysis from a food security perspective*. Luxemburg: European Union, Joint Research Center (JRC), European Commission.

Gaisina, S. (2011). Credit policies for Kazakhstan agriculture. *Central Asian Survey, 30*(2), 257–274.

Government Decree. (1997). *On the procedure and conditions of compulsory agricultural insurance* [in Russian]. Government Decree of the Republic of Kazakhstan No 460 of 2 April 1997.

Government Decree. (2010). *On the approval of the Programme for the Development of Agro-Industrial Complex for 2010–2014* [in Russian]. Government Decree of the Republic of Kazakhstan No 1052 of 12 October 2010.

Government Decree. (2011). On the approval of the Development strategy of "KazAgro National Management Holding" for 2011–2020 [in Russian]. Government Decree of the Republic of Kazakhstan No 52 of 31 January 2011.

Government of Kazakhstan. (1998). *Agriculture, forestry and fishery of Kazakhstan in 1997*. Almaty: National Statistics Agency of the Republic of Kazakhstan.

Government of Kazakhstan. (2018). *Technological modernisation for accelerated growth and better quality of life: Strategic plan for development of Kazakhstan until 2025*. Astana.

Gray, J. (2000). *Kazakhstan: A review of farm restructuring* (World Bank Technical Paper No. 458). World Bank: Washington, DC.

Kraemer, R., Prishchepov, A. V., Müller, D., Kuemmerle, T., Radeloff, V. C., Dara, A., et al. (2015). Long-term agricultural land-cover change and potential for cropland expansion in the former Virgin Lands area of Kazakhstan. *Environmental Research Letters, 10*(5), 1–17.

Land Code. (2003). *Land code of the Republic of Kazakhstan* [in Russian]. No 442-II of 20 June 2003.

Law of the Republic of Kazakhstan. (2001a). *On grain* [in Russian]. Law No 143-II of 19 January 2001.

Law of the Republic of Kazakhstan. (2001b). *On land* [in Russian]. Law No 152-II of 24 January 2001.

Law of the Republic of Kazakhstan. (2004). *Law on mandatory crop insurance* [in Russian]. Law No 533-II of 10 March 2004.

Law of the Republic of Kazakhstan. (2005). *Law on State Regulation of Development of Agriculture and Rural Territories* [in Russian]. Law No 66 of 8 July 2005.

Law of the Republic of Kazakhstan. (2014). *On Introducing Amendments and Additions to Certain Legislative Acts of the Republic of Kazakhstan on taxation matters* [in Russian]. Law No 257-V of 28 November 2014.

Message from the President. (2012). *Message from the President of the Republic of Kazakhstan, Leader of the Nation, NA Nazarbayev: Strategy "Kazakhstan 2050"* [in Russian]. Available at https://strategy2050.kz/ru/page/multilanguage/. Accessed 11 December 2019.

MOA. (2019). *State Program of Development of the Agro-Industrial Complex of the Republic of Kazakhstan for 2017–2021 Ministry of Agriculture of the Republic of Kazakhstan*. Available at www.gov.kz/memleket/entities/moa. Accessed 23 October 2019.

MONE. (2020). *Official statistics in the Agricultural sector statistics Committee of Ministry of National Economy of the Republic of Kazakhstan* [in Russian]. Available at https://www.stat.gov.kz/official/industry/14/statistic/8. Accessed 30 January 2020.

Organisation for Economic Co-operation and Development (OECD). (2013). *Review of agricultural policies: Kazakhstan 2013*. Organisation for Economic Co-operation and Development. Paris: OECD Publishing.

OECD. (2019). *Monitoring the development of agricultural co-operatives in Kazakhstan*. Organisation for Economic Co-operation and Development. Paris: OECD Publishing.

Olcott, M. B. (1981). The collectivisation drive in Kazakhstan. *The Russian Review, 40*(2), 122–142.

OPRK. (2013). *Concept on transition towards Green Economy until 2050*. Office of the President of the Republic of Kazakhstan. Available at https://policy.asiapacificenergy.org/sites/default/files/Concept%20on%20Transition%20towards%20Green%20Economy%20until%202050%20%28EN%2.pdf. Accessed 18 October 2019.

Petrick, M., & Pomfret, R. (2016). *Agricultural policies in Kazakhstan*. Leibniz Institute of Agricultural Development in Transition Economies. Halle (Saale), Germany.

Pomfret, R. (2007). *Distortions to agricultural incentives in Kazakhstan* (Agricultural distortions Working Paper 03, August 2007). University of Adelaide.

Presidential Decree. (1995). *On land* [in Russian]. Decree of the President of the Republic of Kazakhstan No 2717 of 22 December 1995.

Presidential Decree. (2002). *On the State Agriculture and Food Programme of the Republic of Kazakhstan for 2003–2005* [in Russian]. Decree of the President of the Republic of Kazakhstan No 889 of 5 June 2002.

Presidential Decree. (2003). *Programme for Development of Rural Territories of the Republic of Kazakhstan for 2004–2010* [in Russian]. Decree of the President of the Republic of Kazakhstan No 1149 of 10 July 2003.

Presidential Decree. (2010). *On the National Programme of Accelerated Industrial and Innovative Development of the Republic of Kazakhstan for 2010–2014 and Cancellation of Certain Decrees of the President of the Republic of Kazakhstan* [in Russian]. Decree of the President of the Republic of Kazakhstan No 958 of 19 March 2010.

Saparov, E. K., Dauliyeva, G. R., & Tuleshova, G. B. (2018). Clustering of agroindustrial complex of the Republic of Kazakhstan in the context of competitiveness. *Revista Espacios, 39*(18), 16–25.

Skryabin, S. (2011). Subject, method and system of Law in Rem: Some codification issues of civil legislation in Kazakhstan. *Law of Ukraine, 7–8*, 28–41.

Spoor, M. (1999). Agrarian transition in former Soviet Central Asia: A comparative study of Kazakhstan, Kyrgyzstan and Uzbekistan. *ISS Working Paper Series/General Series, 298*, 1–29.

Tax Code. (2008). *Code of the Republic of Kazakhstan "On Taxes and Other Obligatory Payments to the Budget" (Tax Code)* [in Russian]. Code No 99-IV of 10 December 2008.

Tokbergenova, A., Kiyassova, L., & Kairova, S. (2018). Sustainable development agriculture in the Republic of Kazakhstan. *Polish Journal of Environmental Studies, 27*(5), 1923–1933.

World Bank. (1996). *Irrigation and drainage improvement project 1996–2004*. Available at https://projects.worldbank.org/en/projects-operations/project-detail/P008510. Accessed 19 January 2020.

Zholamanova, M., Doszhan, M. A. R., & Kukiev, A. T. A. (2018). Devaluation in Kazakhstan: History, causes, consequences. *European Research Studies Journal, 21*(4), 831–842.

CHAPTER 4

The Role of the Educational Community in Developing Kazakhstan's Higher Education Sector: Autonomy, Self-Organisation, and Capacity to Influence Policy

Dennis Soltys and Aliya Bizhanova

Introduction

The purpose of this chapter is to contextualise and explore the self-organisation of higher education faculty associations in Kazakhstan and their capacity to influence the development of educational policy. The presence of educational associations is a great benefit for educational systems due to their potential contribution, through candid critique and recommendations, to policy and reform (Bascia 2015). They can test new ideas and participate in policy initiatives in pilot modes, so that new programmes can be refined and then scaled up. However, if the educators

D. Soltys (✉)
KIMEP University, Almaty, Kazakhstan
e-mail: dsoltys@kimep.kz; dennissoltys@mail.ru

A. Bizhanova
Almaty Management University, Almaty, Kazakhstan

A. Koulouri and N. Mouraviev (eds.),
Kazakhstan's Developmental Journey,
https://doi.org/10.1007/978-981-15-6899-2_4

who implement curricula on the ground do not have a sense of ownership, the educational system suffers in terms of the quality of the education it provides and the efficiency in the way it operates. In this regard, Frost et al. (2014, p. 235) posit that, "Authentic reform may depend on the extent to which all teachers [and faculty], students and ancillary staff feel a sense of ownership, belonging and moral purposes as members of a learning community". Similarly, Ruby and McLaughlin (2014, p. 298) assert that for the transfer of innovations and reform in Kazakhstan, "a premium [should be placed] on the growth of voluntary professional associations".

In most countries, Western ones in particular, teachers and faculty are the de facto custodians of education (Gappa et al. 2007; cited by Rumyantseva and Caboni 2012). This occurs because societal actors such as students, parents, and employers are fragmented and lack intimate knowledge of educational matters, while state officials (often amateurs) are distant from classrooms. Sources are unanimous in asserting that Kazakhstan's educational system is at all levels excessively centralised and bureaucratic. For example, Tempus (2010) notes that state micro-management puts a brake on change; while the Organisation for Economic Co-operation and Development (OECD) (2018) emphasises the need to strengthen institutional autonomy, particularly in academic, organisational, and financial terms, and thus increase institutional responsibility.

While most faculty unions or associations in developed countries focus on protecting the rights of their members, they also see themselves as catalysts for educational improvement at the institutional level and as representatives of educational interests at the level of national policy. In these countries, faculty constitute a "dignified profession", characterised by adequate salaries, legal protection, opportunities for merit-based professional growth, and respect for their expertise from the side of society and government (Van Keuren 2011). In Kazakhstan, however, schoolteachers are generally considered as technical implementers of centrally defined reforms, not as key players in educational change (Asanova 2007, as cited by Yakavets et al. 2015), while faculty likewise have no incentives for reforms (Sarinzhipov 2017). The academic profession remains poorly paid, for the average base university salary is approximately USD 147 per month for a lecturer with ten years of experience; USD 213 for an associate professor; and, USD 279 for a professor (European Commission 2017). By taking on extra classes and supervising graduate students under the *stavka*—a fee-per-course system that induces faculty to teach many small courses shallowly—faculty (consequently overworked) can increase their income by approximately 25–50% of the base salary. But even in such cases faculty salaries compare poorly with the national average for all occupations of $737 per month (Fimyar et al. 2014). Further, the post-Soviet era approach to education policy development and implementation has been characterised by low governmental trust in teachers and educational institutions (Rumyantseva and Caboni 2012), as evidenced by the existence of a detailed compliance model of quality assurance (OECD 2007) which is oriented (as in the Soviet era) on inputs (e.g. texts and workbooks used, required courses, contact hours, teacher certifications, budget allocations), rather than actual skills and results obtained by students (Yakavets and Dzhadrina 2014; Wilson 2017). The resulting lack of dialogue between educators and state officials reduces the chances to improve educational outcomes (World Bank 2011). Without clear accountability, enhanced by institutional autonomy and empowerment of faculty, the educational system in Kazakhstan is left to the Ministry of Education and Science (MES) officials who, faculty complain, are rarely involved in delivering teaching and seldom speak with those who implement national education strategies.

Accordingly, if teacher and faculty associations could develop their capacity for self-organisation and then for greater participation in and enrichment of educational delivery, this would improve higher education in Kazakhstan. Associations should be encouraged to become a new frontier in higher education development.

Drawing on a mix of reviews of association charters and internal reports, classroom visits and observations, conversations with faculty, interviews with association heads, and the authors' own teaching and field experience, this chapter describes the functioning of three higher education faculty associations in Kazakhstan—the Kazakhstan Association of Teachers of English (KazTEA), the Kazakhstan Association of Young Researchers and Faculty (KAYRF), and the Young Researchers Alliance (YRA)—along with their attainments, their potential, and the limitations in their development. Some attention is also paid to education at the primary and secondary levels, because these levels provide a context for higher education, and because teachers at these levels share with Higher Education Institution (HEI) faculty many of the same systemic challenges in their work.

The next section provides a contextual overview of Kazakhstan's education system, civil engagement in education, and the evolution of higher education. These topics subsume internationalisation, research and development (R&D), governance and quality assurance. This is followed by field studies of three national-level faculty associations, showing the influence of educational associations on higher education policy. The chapter concludes with insights into the challenges of introducing innovations and bringing change to higher education in Kazakhstan, the role that faculty associations can play, the challenges and barriers they face, and how these can be mitigated.

Education Sector: Kazakhstan's Context

An aphorism has it that an educational system reflects the particular kind of society it is called upon to serve (De Witt 1961). In Kazakhstan, primary and secondary education are currently the government's priority, and in a sense set both the institutional and financial contexts for and constraints on higher education. The OECD (2018) has recommended that more attention be paid to primary and secondary education, stronger measures for inclusion of disadvantaged categories of the population, and more effort to reduce poorly achieving segments of primary and secondary education. The OECD's rationale is that primary and secondary education need to be well provided for, if higher education is to achieve its full potential. The Kazakhstani government has been keeping to this advice, allocating, for example, $763 million to build 52 general education schools between 2015 and 2017 (Satubaldina 2018), with 200 new schools planned over the next several years (Expert.gov 2019). The government is strongly financing early childhood education and care (OECD 2018) and is implementing robust policies for ensuring universal access (OECD/World Bank 2015). The unintentional result is that higher education is somewhat in the shadow of reform efforts at lower levels.

Measured by expenditure, Kazakhstan's effort to support its education sector is low by world standards, though tending to increase after a trough between 2004 and 2006 (OECD 2018). State financing of all education was 3.5% of gross domestic product (GDP) in 2017 (OECD 2018), which was a significant increase from an average of about 2.8% between 2009 and 2016 (TheGlobalEconomy.com). Higher education has never received more than 10% of the education budget from a peak in the late 1990s; in 2004, it received 7.9% of the education budget, which represented 0.3% of GDP (OECD 2007). In 2007, Kazakhstan had the world's second-lowest expenditure per higher education student (Wilmoth 2011). By 2015, public expenditure on higher and postgraduate education had reached 0.5% of GDP (European Commission 2017). Overall, the government's spending on all levels of education combined still lags behind the international benchmark of 5.6% of GDP (OECD/World Bank 2015).

Despite low spending, the MES has achieved much since national independence. It has re-constructed or constructed many parts of the governance and infrastructure in education and has established a robust system of monitoring, licensing, and quality control in education at all levels (IQAA 2016). Trilingual instruction (Kazakh, English, and Russian) is being implemented across the country at the primary level (Mehisto et al. 2014), while at the secondary level instruction is being extended from age 11 to 12 years (Fimyar et al. 2014). Supportive rhetoric and practical assistance are given to inclusion and student-centred learning. Accordingly, more children with disabilities are being integrated into regular classrooms. Moreover, teachers and faculty are beginning to teach in a less authoritarian manner (Rouse et al. 2014). A sizeable system of Centres of Excellence for upskilling teachers has been set in place at the primary and secondary levels, constituting a "cascading" system of teacher upskilling whereby newly trained cohorts of teachers train additional cohorts (Turner et al. 2014). This is matched by a National Centre for Professional Development for upskilling university faculty (IQAA 2016).

In addition, the coverage of small rural schools and pre-school facilities is being increased after a decline in the 1990s (Yakavets 2014). New mechanisms of education financing are being tested and implemented, including more private financing at the local level as well as a fairer formula of per-capita state financing in secondary education (Bridges and Sagintayeva 2014). Financial grants are given to indigent rural applicants to attend higher education after individual testing. However, secondary schools do not have systematic guidelines to support indigent students who are falling behind in their studies (OECD/World Bank 2015). A total of 16 secondary "lighthouse" Nazarbayev Intellectual Schools have been established in each region of the country. These are disproportionately well financed at a per-student rate of three times the national average (OECD/World Bank 2015).

Overall, educational opportunity and quality in rural areas still lag considerably behind the cities (Fimyar 2014); while the cities, in turn, lag behind international quality standards (OECD/World Bank 2015). According to the Programme for International Student Assessment (PISA), in which Kazakhstan has participated since 2009, the country's 15-year-olds are on average two years behind their peers from OECD countries (OECD/World Bank 2015). Similarly, the World Bank (2017) confirmed a growing gap of student achievement between rural and urban schools. This presents a burden and environmental constraint for HEIs, which receive entry cohorts that are under-prepared in terms of knowledge, skills, and aptitudes (OECD/World Bank 2015). Quality in education is currently one of the MES's main concerns; and, to the Ministry's credit, Kazakhstan's education level has risen somewhat in international rankings such as Trends in International Mathematics and Science Study (TIMSS) and Progress in International Reading Literacy Study (PEARLS) (Information-Analytic Centre [IAC] 2018). In 2018, the quality of Kazakhstan's higher and vocational education was ranked at a respectable 56 out of the 137 countries measured (Seedstars.com 2020).

Civic Engagement in Kazakhstan's Education Sector

Soviet culture discouraged the development of civic associations. Despite this, while reforming the governance and infrastructure of the nation's education sector, Kazakhstan's leadership gradually realised that non-governmental organisations (NGOs), and professional and civic associations could contribute specialist knowledge, policy ideas, and energy to improve policy design and implementation (Soltys and Orynbassarova 2013). The 2000s saw a series of laws designed to encourage civic associations to form, and from 2001 the national government has sponsored a biannual Civic Forum in which civic associations, NGOs, governmental officials, and other stakeholders could discuss policy ideas. Education groups have been present in these forums and have become part of a greater movement in Kazakhstan for local groups to be at the forefront in the evolution of civil society. In appreciation of good works, the government has provided facilities and seed money for civic and professional associations and intends to strengthen them through the mechanism of Social Contracts (Soltys 2014).

In the realm of higher education specifically, state policymakers have concluded that a system based on greater institutional autonomy and accountability, similar to those of some European countries and the USA, offers the most promising strategy for improving quality (Sagintayeva et al. 2017). The need for institutional autonomy follows from the increasing tendency for higher education to be more competency-driven than input-driven, and from the related fact that the student body (often having many members with prior professional experience) is more diverse. That is, institutions must have sufficient flexibility to arrange their programmes so as to meet the diversity and complexity of the student body (Plessius and Ravesteyn 2010). However, teacher and faculty associations in Kazakhstan have low engagement with outside stakeholders (OECD 2018), and, throughout national independence, NGOs and state agencies have generally underperformed in engaging with local communities (Soltys 2015).

Habits of "command and control" remain hard to break, and excessive state oversight continues with several drawbacks. First, the existing system of teaching and testing is still focused too much on "what to teach" instead of "how to learn" (Yakavets and Dzhadrina 2014). Similarly, assessment is rarely seen as a formative reflection on practice or study (Winter et al. 2014). Second, the MES emphasises accountability for inputs without giving credit for value added. Further, instructions from the distant capital are not always appropriate for the conditions of specific schools and classrooms, which do not fit notions pre-conceived at the centre and are inappropriate for diverse student cohorts which should not be treated in a uniform manner. Consequently, monitoring and measuring by the MES are sometimes conceptually misguided—the wrong things are measured and/or are measured incorrectly, resulting in design and implementation of policies that do not achieve their objectives (Yedgenov 2019).

Bascia (2015) notes that pedagogical associations play an important role in defending public education and controlling the damaging effects of ill-thought-out educational policies. "Command and control" methods are often difficult or even impossible to implement, despite administrative pressures. Moreover, instructions from the centre can be either vague or self-contradictory, resulting in educators' being left without practical guidance or support for implementation. Such an example is the common MES desire to reduce student attrition, but also to maintain high academic standards. Another example is the desire to upgrade the teaching skills of faculty through time-consuming, mandatory re-training, at the same time that the MES desires to retain faculty who are active in research and getting their work published. In such cases, HEIs must find workable, ad hoc, compromises between conflicting demands.

Such pathologies could be mitigated if faculty were trusted and resourced to implement tasks themselves. But in the absence of freedom of discretion at the ground level, administrative methods soon come up against the so-called "control paradox" (Raza 2009), namely that the more a state agency tries to bureaucratically control institutional processes, the more unworkable and weaker the controls actually become. As Sen (2000, p. 3) noted, "responsibility requires freedom", and in the absence of sufficient freedom, education controls and policies underperform.

Perhaps more controversially, many of the government's actions seem to be motivated by a narrow economic, human capital rationale (Fimyar et al. 2014; Shamshidinova et al. 2014) and a race for global rankings; possibly undervaluing the humanistic and social sciences sides of education. The authors' experience of university-level Advisory Board meetings with employers shows that employers share a view held by wider society that education should be "practical", without being aware that a liberal arts component makes graduates more productive within the economy. Graduates from the liberal arts-oriented KIMEP University, in Almaty, are strongly preferred by employers because of the "soft" and transferrable skills that they possess, as evidenced by an employment rate of over 90% within six months of graduation (KIMEP University 2019), as opposed to a national average of 80% for masters students after one year (European Commission 2017).

In view of the above, Kazakhstan in 2019 still does not have a general and higher education community with robust channels of communication with the MES or the capacity to lobby effectively for education interests at the national level. As a result, the development of the education sector is left to the discretion (or, in some cases, neglect) of state officials who may feel that other policy needs are more pressing. Anderson (2006) noted correctly that in the Soviet era there were no strong public lobbies for spending on health and education, which were provided for at the discretion of the executive branch of the government. This appears to still be the case: currently, hierarchical education policymaking in the country inhibits the participation of civil society and interest groups (OECD/World Bank 2015). The experience of some Western countries, such as Canada, the USA, and the Scandinavian countries shows that education systems are strong when teacher and faculty associations have regularised communication with state education officials (Nusche et al. 2016). In some countries such as Denmark, Canada, Ireland, and Finland, teacher and faculty unions have a long history of identity formation and participation in education policymaking, including lobbying of national governments for teacher and faculty interests (Carr and Beckett 2018). In addition to these activities, educator associations can influence education policies for the better by providing good-quality empirical research, and trading this expertise for admittance to evidence-based policymaking.

Evolution of Higher Education in Kazakhstan

Under Soviet rule, from 1920, state-sanctioned teacher associations in Kazakhstan were established, in the terminology of the time, to serve as policy "transmission belts" from the state leadership to the pedagogical rank and file. Vocational and technical education was designated along functional lines for service to industry (Soltys 1997), while tertiary education was also extensively vocationalised (Heyneman 2008). Such a model caused at least two overriding problems that remain to the present: vertical fragmentation and compartmentalisation within the administration and delivery of education, and a lack of involvement in education matters from societal and economic interests.

After the country gained its independence in 1991, Kazakhstan's MES continued to operate in the old spirit, and the working culture was one of conformity rather than of elicitation of constructive criticism of policies by educators and the wider civic society. At post-Soviet "ground zero" for educator associations, it was inevitable that, for a time at least, the MES and its agencies alone would fill the institutional void in the education realm, and that education policy would take on a top-down character. Ahn et al. (2018) note that although the philosophical underpinnings of Soviet higher education for a managed economy were uprooted, the "ownership" of education did not change; it remained with state officials. The 1990s were a time of financial contraction and institutional disorganisation. Consequently, much of the decade was taken up simply with establishing a normative and legal base for the education sector (Massyrova et al. 2015). The government improved the political environment by passing a law on NGOs in 2001 (Law of the Republic of Kazakhstan 2001), which defined all teacher and faculty associations as non-profit entities. In 2007, the National Law on Education (Law of the Republic of Kazakhstan 2007, pp. 13–14) stipulated the "democratic nature of education management and [the] principle of transparency of the education system", though it did not prescribe workable procedures through which community groups and educators could add their voices to policy development and implementation.

Ahn et al. (2018) offer a convenient taxonomy of the main changes in Kazakhstan's higher education, dividing them into three generic categories: internationalisation, R&D, and organisation and quality. These occurred alongside three other, subsidiary, changes, namely horizontal diversification, vertical diversification, and inter-organisational relationships.

Internationalisation

Regarding the first of the generic changes, governmental authorities were quick to appreciate the benefits of international contacts. Thus, from the early 1990s the MES welcomed university-level international partnerships and experiments. Some of these include the private KIMEP University, established in 1992 as a liberal arts, North American-style institution teaching in English; the Kazakh-British Technical University; the Suleiman Demirel University; the Kazakh-American University; and the Kazakh-German University. All this was in keeping with President Nursultan Nazarbayev's vision, articulated publicly at least as early as 2006 (Nazarbayev 2006), that Kazakhstan would not cower in isolation from global trends, but would boldly ride the crest of globalisation by investing generously in education and science. The country would be open to the free flow of ideas, people, and capital. Striking evidence of this commitment is the *Bolashak* (Future) programme of financial assistance for university students wishing to study abroad. Since its inception in 1993, the *Bolashak* programme has supported over 12,831 students (IAC 2018).

Meanwhile, 11 leading universities and the new Nazarbayev University, in the country's capital city, have been designated as regional research hubs. Nazarbayev University is generously financed, and aspires to become a "world-class" institution (Katsu 2017). Moreover, in 2019, Al-Farabi Kazakh National University increased its standing in the Quacquarelli-Symonds World University Rankings to reach 220th place, Satbayev University was placed 464th, and Auezov South Kazakhstan

State University 480th; that is, three Kazakhstani universities were placed within the top 500 world universities (Quacquarelli-Symonds 2019).

A major landmark in internationalisation and impetus for internal reforms was membership in the European Higher Education Area (EHEA) in 2010, along with accession to the Bologna Process the same year (Ahn et al. 2018). This led the MES to introduce a large number of upstream institutional changes and innovations. One of these was the adoption of the European Credit Transfer and Accumulation System (ECTS), which made Kazakhstani academic degrees recognisable by European and other institutions. The adoption of the ECTS required the adoption of the Dublin Descriptors (European Commission 2015), which in turn required deep re-thinking of the adopted pedagogical philosophy, making it more student-centred and outcomes-oriented. This also required better data collection and record-keeping, so that adherence to international standards could be proven to domestic and foreign accreditation agencies.

Research and Development

A horizontal change was intended to be contractual links with outsiders in the area of R&D, but results here have been very modest. Kazakhstan spent only 0.25% of GDP on R&D in 2007, and this was mainly by the state (OECD 2007). Research and innovation activities are historically the least reformed area of the Kazakhstani higher education system. The modest results in outsider involvement are attributed largely to the compartmentalisation of academic research institutes, universities, faculties, and domestic and foreign markets (OECD 2007).

Governance and Quality Assurance

The second and third major changes were in the areas of organisation and quality assurance and involved privatisation and stratification.

Innovations in education governance were introduced through a notable horizontal change: the introduction of *Boards of Trustees* or Supervisory Boards for secondary and higher institutions. New Boards of Trustees were set up between 2012 and 2016 in 19 state universities, and given authority over the allocation of sponsorships, charitable assistance, and funds received from non-governmental sources. The boards were further given authority over the appointment of rectors, previously done by the government. By 2018, over half of the country's universities had established governing boards (OECD 2018), the establishment of which was permitted to universities that had achieved international accreditations.

Another significant horizontal change was the *privatisation of HEIs*. As of 2015, there were 55 public and 70 private universities in the country, enrolling 247,000 and 259,000 students, respectively (European Commission 2017). Between 40 and 50% of the revenue of public HEIs is generated through paid educational services (largely tuitions), while private sector enterprises contribute only 0.9% and regional and local authorities 1.4% (European Commission 2017). Private universities, on the other hand, generate at least 70% of their revenue from educational services, with private sector enterprises contributing 2% and local authorities 0.9%. The majority of contributions by regional authorities are not discretionary but are required by central government, and therefore cannot be properly considered as regional (OECD/World Bank 2015). Income generated from research and innovation in private HEIs represents only between 0.1 and 3.2% of the generated revenue (European Commission 2017), illustrating the low level of HEI links to stakeholders. It is evident from these figures that privatisation has not led to a significant diversification of income sources.

The country's tertiary education is in general highly privatised, with 85% of students being self-funded and 14.7% on state grants (Expert.gov 2018). If public and private spending is counted together, Kazakhstan spends 1.5% of its GDP on higher education (OECD 2018). In 2014, 70% of this expenditure came from funds generated by private sources (tuition fees); a significantly higher percentage than the OECD average of 30% (OECD 2018).

Though not financed by the government, private HEIs must hold to the MES rules and standards. Privatisation allowed many HEIs of dubious quality and ethics to spring up, with the number of such institutions reaching a peak of 181–182 between 2001 and 2005. However, in the light of the need for regulatory and quality control, the MES induced some of the marginal ones to close or merge with others, leaving 127 public and private institutions in 2015 (IQAA 2016; Sagintayeva et al. 2017). In fairness to the MES, it should be acknowledged that many of these institutions did indeed require regulation. In other words, both administrative voids and the emergence of institutions of dubious standards seemed to confirm the necessity for strong MES leadership.

A further important change identified by Ahn et al. (2018) is the vertical *stratification* of HEIs, motivated in large part by a desire to rationalise and concentrate R&D in the most viable institutions. Accordingly, legislation currently provides for seven kinds of HEIs, listed in rough order of decreasing scope of autonomy (IQAA 2016; European Commission 2017). These are:

1. 1 Autonomous Organisation of Education (AOE) (Nazarbayev National University)
2. 11 National research universities
3. Research universities
4. Universities
5. Academies
6. National higher education institutes
7. Institutes.

HEIs are differentiated according to the breadth of their programme offerings and depth (degree levels) of instruction, and also by their governance structure. National research universities are allowed to set their own tuition fees and faculty salaries, have Consultancy Boards or Boards of Trustees, and hire their own rectors. Academies and institutes are designated for special purposes, such as pedagogical training or the arts. As noted above, there is one AOE, namely Nazarbayev National University, in the capital city. Autonomous HEIs are allowed academic freedom for 55% of the content of bachelor programmes, 70% of master programmes, and 90% of doctoral studies (IQAA 2016). Thus, MES regulation is receding for top institutions, but remains significant at the lower levels.

Other Regulations

The list below provides examples of regulations of little or questionable pedagogical value. These include the following:

- The MES may demand the inclusion of courses that do not fit well into academic programmes and make the programmes longer and more expensive for students.

- There must be two mid-term assessments of students totalling precisely 30% plus 30% of the final grade, which can be pedagogically somewhat inappropriate and/or unsuitable.
- At least one of the main textbooks for a course must be no older than five years. This is inappropriate for courses that are based on classic works and also assumes that recent books are of better quality than older ones.
- Graduate students are required to publish a piece of work prior to the completion and defence of their thesis. The pedagogical merit of publishing academically non-vetted pieces of work seems questionable. Moreover, adherence to the requirement necessitates—at a time of stress for the students—the search for a suitable publication outlet, or the hasty organisation of a student conference in whose proceedings the inclusion of work will count as a "publication".
- Faculty who supervise master theses must have at least ten publications in the last five years, whereas faculty who supervise doctoral work must have 20. These publications must include and least one teaching textbook or equivalent, and Pedagogical Guides for the courses taught. These specifications are difficult to achieve for many faculty, especially considering the heavy teaching loads of approximately 750–1100 contact hours per year (European Commission 2017). The rule specifies that only one of those publications needs to be in a higher-level international outlet, while the others presumably may be smaller or lower quality pieces. As a result of overwork, quantitative controls, and lack of monetary or other incentives, faculty are perversely induced to publish a large amount of low-quality work.

Such measures compound the time consumed by the detailed reporting necessary for frequent national and international attestations and licences. A reduction in teaching loads is a high priority for faculty members. However, the absence of measures to address high teaching loads indicates that the MES might be simply ignoring this issue and constitutes evidence that faculty associations have been unsuccessful in negotiating with the MES on an issue that is important for their members as individuals and for the quality of educational delivery more broadly.

The Emergence of Faculty Associations

This section provides an overview of three higher education faculty associations operating nationally, and which, as of 2019, are the only ones to have come to the fore and be registered as non-profit entities in the National Business Archive.

Based on a wide array of information, field experience, and interviews, as mentioned, this study looks into the goals and current projects of these national-level associations, and the challenges they face in carrying out their activities in the higher education arena. Representatives from each of the three associations were interviewed about their associations' role in policy initiatives, their contribution to educational reforms, and their vision for further development. They were also asked about their associations' strengths and weaknesses, and the obstacles their associations face. They shared their opinions about the potential for collaboration between central government authorities and the associations, and the reasons behind the reluctance of education policymakers from the centre to do more to encourage participatory quality assurance systems to provide feedback on reforms.

Case 1: The Kazakhstan Association of Teachers of English (KazTEA)

KazTEA is a union of professional organisations from seven regional cities of Kazakhstan, namely Aktobe, Karaganda, Kostanai, Taraz, Kyzylorda, Shymkent, and Uralsk. Its membership of around 1500 includes mainly university faculty and teachers who specialise in the English language, but also those who teach other subjects such as chemistry, maths, science, or literature in English. KazTEA does not have its own office but uses schools and university venues for meetings. Funding is scarce, but some international and national partners, such as the American and British embassies, and other foundations, offer occasional grants.

Its mission is to increase the professional capacity of English-language faculty, develop cooperation among its members and beyond in research related to linguistics and English-language development, and to exchange knowledge at the national and international levels. In addition, its charter includes support of the members' professional interests at the policy level and collaboration with the MES and local education authorities.

The association partners with local state education agencies and state departments for youth development, and focuses on several small school and HEI language initiatives related to English-language promotion. Other partners are the Republican Institute of Teacher Professional Development, Nazarbayev Intellectual Schools, Nazarbayev University, the Regional Office of English Language at the US Embassy, the British Council, and the Macmillan and Oxford University Press publishing houses. KazTEA's engagement with central and local education authorities is limited.

Key activities include research- and practice-based conferences, training events, language summer camps, and round tables; in particular:

- Core members contributed to the development of a unified occupational standard for language teachers, which has been accepted by the national government.
- British Council projects enable collaboration with UK institutions. KazTEA members train in language-teaching methodologies and develop methodological guidelines. As a result, training materials for interactive teaching of the English language, project management, and pedagogy have been developed by KazTEA members, and serve as supplemental materials in schools. Also, through collaboration with the Cambridge Test Centre, members study for the Teaching Knowledge Test.
- University English-language faculty develop trilingual education methodologies at their institutions, organise classes, and author course materials focused on innovative methods of teaching the English language.
- The Association's branch in Shymkent administers an English Access Micro-scholarship programme that provides English-language courses to teenagers from socially vulnerable backgrounds.
- The Hornby Education Trust Fund from the UK sponsors the Association's training of rural schoolteachers in the use of interactive teaching methods, learning strategies, and assessment.
- Some members participate in professional development through the Fulbright programme, by interning and conducting research in US universities. Others participate in Erasmus Mundus, Tempus, and Erasmus + programmes, and conduct joint research on linguistics with European colleagues.
- The KazTEA takes part in international conferences on topics of linguistics, English-language teaching (TESOL), pedagogy, and interactive education. Over 150 members conduct online discussions

and seminars on topics related to critical thinking in English-language teaching, assessment, online project-based learning, and teaching methodologies.

In view of its activities, the KazTEA has done much practical groundwork in the areas of teaching methodology, curriculum design, teacher training, and standards. It has a broad interface with both domestic and international organisations. However, with regard to policy-related work, Association leaders have expressed concern that communication with the MES and local education authorities is currently one way, contributing to the implementation of a top-down education agenda. The KazTEA is not asked to provide feedback on the implementation of ongoing reforms. The Association consequently provides minor reports that are not focused on evaluation, but rather on inputs such as the number of teachers trained and of seminars conducted. As one of the Association leaders stated in an interview:

> The MES dictates to us what to implement; it is one-way, top-down communication. This is not effective. We have not received any inquiries from the MES about our opinion of the ongoing implementation of new curricula nor evaluation results of implementation. We regularly send information letters about our events to the MES, and that is all. In order to start a dialogue, we might need to conduct a comprehensive analysis of the existing situation with educational associations and the role they see themselves playing in this dialogue. Owing to financial struggles and no external support, such analysis has not been initiated.

The KazTEA sees itself as independent and non-biased, and its leaders think that the MES is not open to receiving critical feedback or recommendations for policy improvement. The Association leadership's aspiration is to conduct rigorous research related to linguistics, rather than merely to organise events called for by the MES. Moreover, the Association receives much feedback on reform implementation from its members and would like to build its capacity for transmitting this information to national decision-makers. At the moment, KazTEA's leadership does not see effective mechanisms for participation in policymaking and feels itself actively discouraged from sharing viewpoints or research results that might imply criticism of current affairs.

Case 2: The Kazakhstan Association of Young Researchers and Faculty (KAYRF)

The KAYRF was established in 2018, initiated by the Kazakh National Women's Pedagogical University, in Almaty, as a result of an Education and Science Minister's recommendation. It is a voluntary professional organisation for tertiary-level faculty and researchers, and has approximately 1000 members. Currently, the Association includes participants from 20 national and local HEIs focused on pedagogy and developing faculty under the age of 40. Work has been started to include secondary schools, but the number of members from schools is small so far. The KAYRF does not receive funding from the MES or other sources; hence it relies on in-kind assistance such as facilities and office space from member universities. Its mission is to unite innovative faculty and researchers who contribute to the development of the state and society, and to boost scientific, practice-based research on, and exchange of, pedagogical knowledge. The KAYRF aims to create a discussion platform for young researchers in Kazakhstan and to develop sustainable partnerships with international educational institutions, unions, and schools (Kazakhstan Association of Young Researchers and Faculty 2018). The members strive to be socially responsible teachers and faculty. The Association's founding members consider their work as groundbreaking, because it seeks to unite teachers and faculty of various pedagogical education institutions around the country within a structure that has not existed before.

The KAYRF's core functions are to contribute to improving the research capacity of its members, and thus to participate in educational events related to pedagogy, teacher professional development, curriculum development, and education policy discussions at the national level. During its short operational life of about a year, the Association led or participated in a number of activities, such as:

- The development of online courses for the OpenU.kz project, which is government-financed under the *Rukhani Zhangyru* (Spiritual Renovation) programme. This government-sponsored programme aims to develop textbooks by translating the newest and most popular books into the Kazakh language, to be used in schools and HEIs.

- A recently initiated partnership with local schools in Almaty city. Members conduct research seminars for schoolteachers and administrators, focusing on pedagogy and trilingual education.
- Training events (jointly with British colleagues) on "action research" focusing on diagnostic analysis for the continuous improvement of practices and policies. Current research projects focus on trilingual education and rural school improvement.
- Teacher-to-Teacher Networking (T2T Networking) initiated in 2018 allows for collaboration of pedagogical university faculty and graduates, teachers, school administrators, local government officials, and university students who are soon to be teachers in rural schools. The various groups exchange ideas and experiences and help develop their professional skills.
- Cooperation with United Nations Educational, Scientific and Cultural Organisation (UNESCO) in girls' education, assisting in staging events such as the International Women's Day and the International Day of Women Scientists. Moreover, in partnership with UNESCO, an International Education Day was organised to foster civic understanding of the various purposes of education.
- Leadership of a number of roundtables and training events on applied topics, such as how to publish a paper or how to conform with anti-plagiarism rules.
- Contribution to an MES meeting with young researchers to discuss topics such as graduate and postgraduate education reform, research activities, publishing, and improvement of the science sector in the country.
- Initial steps to participate in a dialogue with the MES on a draft of the Law on Teachers' Status (passed on 27 December 2019) which intended to reduce the administrative and reporting tasks of teachers, enabling them to focus on teaching and teaching-related tasks. KAYRF provided a number of recommendations related to the retirement age and salary system, but members were not sure whether these were accepted, as no feedback has been provided by the MES.

Although the KAYRF does not receive any organisational financing from the MES, it adopts a supporting position concerning MES policies and operations. Until now, it has not seen itself so much as a catalyst for change or candid feedback, but rather as a contributor to ongoing programmes that the MES and its subordinate agencies initiate. The KAYRF has not yet received any requests from the MES or other government agencies to contribute to policy-related domains. Most of the Association's activities have been focused on the preparation of educational events, though the first steps in policy planning and implementation have been initiated.

The KAYRF currently does not have a strategic plan or multi-year vision, but intends to draft one in early 2020, after consultations with its members and university administrators. The Association's founding members expressed the goal to create sustainable research partnerships in broad areas of pedagogy among members and represented universities. Consequently, the future focus will not be primarily on organising joint events or seminars led by the MES, but on policy research, evidence-based reform proposals, publications, commercialisation of joint projects, and cross-institutional collaboration. One of the Association's leaders explained:

> We are still a very young association that was established a year ago and are in the process of self-identification. We face such issues as financing (as most funds come from one university) and lack of time of our members. Most of them are teaching faculty, so they are busy with teaching and paperwork and can rarely participate in policy-level dialogue. The capacity to conduct analysis and research on topics the MES deems important by our members is currently limited. Therefore, we would like to strengthen the research capacity of our members and initiate this kind of two-way cooperation. As we gradually grow, we plan to contribute to the policy-making processes in 2020 more through participation in MES roundtables and discussions related to the teaching profession, pedagogy, and research.

The KAYRF aspires to join international educational associations such as the European Education Research Association (EERA) and plans to participate more actively in education policymaking through regular dialogue with the MES. It is in the process of developing a website and online platform for information exchange between members on a permanent basis. Currently it has limited lobbying power and public support and needs more time to define its role in education policymaking at the national level. The KAYRF is one of the youngest and the most ambitious associations of young teachers and faculty in the country to date.

Case 3: The Young Researchers Alliance (YRA)

The YRA is another young association in higher education. It is a self-managed, non-profit association established in 2018 by young and early-career researchers and students from various fields. Most of the founding members studied for their masters or PhD degrees abroad between 2015 and 2018 and then started discussions on creating an association. At the beginning, its members were mostly from Nazarbayev University, the flagship HEI in Kazakhstan. Currently, most of the 200 + members are higher education faculty and graduate students under the age of 40, who reside in the largest cities, such as Almaty and Nur-Sultan. Some 30% of the current members are Nazarbayev University faculty, while the rest come from other higher education and research institutions. The YRA is operated by its members on a voluntary basis, with the only funding currently received being for graduate student grants from Nazarbayev University (see point list below).

The YRA aims to support young researchers, provide networking opportunities to them and to those interested in research and science, and popularise science among the general public. Its vision is "to create a supportive environment for a new generation of researchers, scientists and academics through effective communication and knowledge exchange among its members and other research-related activities of young researchers" (Young Researchers Alliance 2018, p. 2). Its motto is "Unite. Share. Explore." Apart from research goals, the Alliance aims to provide a platform to discuss scientific and social issues of young researchers, and to contribute to policy-level document formulation and assessment related to the interests of young faculty and researchers.

The Alliance conducts a number of community- and policy-level events and undertakes activities including:

- *Gylym Tanu* (sharing knowledge) monthly workshops that provide an informal platform to talk about research, allowing participants to be acquainted with various research fields and people working in research and science. At each session, two researchers share their research projects and results with the general public.
- A more formal forum, held annually with the participation of researchers from various fields, business representatives, and policy-makers. Its purpose is to connect business and policy to research and science by permitting researchers to share their research results and

best commercialisation practice, and by permitting policymakers to meet researchers who could potentially contribute to policy planning in their fields.
- Soliciting funds for research projects from the government and business sectors, to help sustain the research careers of Alliance's members and provide evidence-based solutions to social and economic issues. Alliance's leaders note that many policy decisions are made without sound evidence gathered from research and aim to persuade government authorities to involve researchers and scientists in policy planning and analysis.

The YRA also provides grants to graduate students interested in conducting research. As of now, MES research funding is provided only to Nazarbayev University students, but the YRA is looking for funds to expand the programme to other institutions. Alliance's leaders see this grants programme as an alternative to the MES research grants competition, the fairness of which they have concerns. Consequently, alliance grant applications go through an open competition, a two-phased blind peer-review, and a careful selection process. One of the YRA leaders related in an interview that:

> We realise that the current distribution of research grants conducted by the MES does not work. Through our pilot programme we strive to create a new, clear mechanism of selecting and awarding research grants. We aim to establish a list or register of experts to create a fair system of research grant evaluation, and hope that the pilot results will be used by the MES in how it distributes research funding.

Moreover, the YRA's leaders became members of the MES Council of Young Researchers, formally attached to the MES, which was initiated in 2019 to start a dialogue with the MES on issues related to education and science in the country. Many members contributed to the dialogue, by conducting analyses of existing issues in science and providing a number of policy recommendations. However, they reported that some MES staff in charge of implementing strategic policy decisions do not share the same vision and, as a consequence, do not deliver what has been agreed at the leadership level. This led some of the members who joined the council in 2018 to resign a year later, feeling that there were no tangible outcomes from collaborating. These members outlined problems such as the lack of meaningful communication with the MES, the inability of the MES to take the council's recommendations on board, and the non-implementation of approved recommendations due to MES bureaucracy and staff changes. Nonetheless, given the YRA's visibility, members are now being invited to various governmental working groups related to the development of research and science in different sectors.

The Alliance is young, although it has already proved to be active and effective in engaging and communicating with the MES and other stakeholders. It considers innovative ways to contribute to policymaking, through developing a list of research experts in various areas and stimulating awareness among the general public about the need for research. The YRA believes that some policy-related issues go beyond the MES's scope of duties, and plans to expand its collaboration with other governmental agencies.

Table 4.1 summarises the key characteristics of the three faculty associations discussed above, including their establishment dates, membership sizes, geographical coverage, stated missions and key activities, and engagement with the MES and other stakeholders.

The Influence of Faculty Associations in Higher Education Policy

Faculty associations' strength lies in their idealism and ability to self-organise. Underfinanced, in most cases lacking administrative support, and based just on the personal motivation of volunteers, they have been able to attract considerable numbers of teachers and faculty across Kazakhstan. Involvement in the activities of associations inevitably means extra work and incidental expenses, without compensatory pay or recognition. Many members complain that they are not able to make meaningful contributions to research or evidence-based policy development owing to heavy teaching workloads. Overwork is therefore probably the main impediment to expanding the membership of faculty associations.

A serious additional impediment is overly bureaucratic state administration, which does not allow faculty sufficient discretion to implement and adjust new MES curricula. Faculty are obliged to follow the instructions provided by the MES, as they are checked for programme licences and periodic attestations, which can go so far as withdrawing programme licences. Faculty note that external stakeholders, such as business representatives or NGOs, are interested in collaboration, but insufficient

Table 4.1 Summary of association characteristics

	KazTEA	*KAYRF*	*YRA*
Establishment date	1999	2018	2018
Membership size	1500	1000	200
Membership type	Faculty and teachers specialising in English language and/or teaching in English	Tertiary-level faculty and researchers; small number of secondary-level teachers	Tertiary-level young/early-career faculty and graduate students of various fields
Geographical coverage	Aktobe, Karaganda, Kostanai, Taraz, Kyzylorda, Shymkent, Uralsk	All major cities; 20 pedagogical HEIs	Nur-Sultan, Almaty; 30% of members from Nazarbayev University, remaining from other higher education and research institutions

	KazTEA	*KAYRF*	*YRA*
Stated mission	• Increase the professional capacity of English-language Faculty • Develop cooperation among its members and beyond in research related to linguistics and English-language development • Exchange knowledge at the national and international levels • Advocate for members' professional interests	• Unite innovative faculty/researchers contributing to developing state and society • Boost scientific, practice-based research and pedagogical knowledge exchange. • Create discussion platform for young researchers • Develop sustainable partnerships with international education institutions, unions and schools	• Support young researchers through effective communication, knowledge exchange and research-related activities • Provide networking opportunities to young researchers and to those interested in research and science • Popularise science among the general public. • Provide a discussion platform for scientific and social issues • Support research to underpin evidence-based policymaking • Contribute to policy-level document formulation and assessment

(continued)

Table 4.1 (continued)

	KazTEA	*KAYRF*	*YRA*
Key activities	• Organise research-and practice-based conferences, training, summer camps, and round tables • Develop unified occupational standard • Author course materials for trilingual education • Provide training of rural teachers • Participate in projects with the British embassy and in Fulbright programmes	• Improve research capacity of members through training events, professional development activities, curriculum development and education days • T2T Networking • Develop online courses for OpenU.kz • Develop partnerships with local secondary schools • Contribute to education policy discussions nationally • Dialogue with MES on policy and legislation	• Popularise science and research through public and community events • Identify issues related to science and research for young and new researchers • Solicit funds for research projects from the government and businesses to provide evidence to inform policy
Funding	Occasional grants from national and international partners	No state funding.	No state funding; seeking funding to provide grants beyond Nazarbayev University graduates
Engagement with MES and local/regional state education agencies/departments	Limited with some engagement with local state education agencies and departments regarding youth development; one-way communication with reports from KazTEA to MES	Provides support for MES reforms; contributed to the discussions of draft Law on Teachers' Status. Limited lobbying power.	Members participate in the MES' Young Researchers' Council and provide recommendations for policy improvement in relation to the science sector. Members invited to various government working groups for the development of science and research in various sectors

	KazTEA	*KAYRF*	*YRA*
Engagement with NGOs, business, and the wider civil society	Schools, US and British embassies, English language centres, international publishing houses	Schools, universities, local government education authorities, UNESCO	Universities, general public interested in science

communication channels have been established under the auspices of the MES or regional education authorities. In view of the traditional lack of public support in Soviet and post-Soviet education, the YRA has astutely opted—through its public engagement activities—for the "long cure" of first trying to interest and then to bring on side the general public.

It is difficult to judge the impact that teacher and faculty associations have made on official policymaking. Policymaking is still tradition-bound to a significant extent and concerned primarily with providing inputs and checking compliance. But even so, official policymaking does not occur in a vacuum. It is therefore likely that at least some of the intellectual contributions and practical suggestions of educators have percolated upwards and have been absorbed, informally, into policymaking. At the very least, as noted, a considerable number of teacher- and faculty-produced workbooks and standards have been adopted and used by the MES. Furthermore, the YRA seems to have established effective communication channels with some MES officials.

In the meantime, the result is a vicious circle of overwork, insufficient opportunities for input, and ministry domination. Faculty associations face several challenges in expanding their geographical coverage, broadening programme operations, and exercising policy influence. The MES and its agencies have been, at times, both helpful and indifferent. Despite good intentions and, in some cases, fruitful collaboration, faculty–ministry relations have underperformed.

The first reason that can be posited for underperformance on the ministry side is structural governance constraints. Because the MES operates under a strictly defined legal and financial public administration framework, it is difficult to introduce innovations that are not within the MES's legal mandate. The budget of the unitary state is highly centralised, consequently municipal governments (*maslikhats*) are allowed little discretion over funds for local capacity-building; and local initiatives distract central agencies from their executive obligations. MES officials therefore remain focused mostly on reporting on inputs to higher bodies in the government, such as the Presidential Administration, the Prime Minister's Office, or the Parliament, rather than on actual problems.

The second, related reason for underperformance is regulatory constraints. There is a lack of clear procedures for dialogue. A rather subtle point here is that governmental agencies need to be designed as more open "learning organisations". That is, public service personnel should function more as receptors and enablers of community initiatives than as regulators and controllers, and should be given greater scope and resources to do so. Meanwhile the MES faces human capacity challenges in not being able to attract qualified personnel, because the national public service is in general underpaid and the turnover is high (Emrich-Bakenova 2009; Emrich-Bakenova 2019). As a result, MES staff do not accumulate sufficient experience in implementing changes to educational governance. Stipulating the role of faculty associations and establishing clearer mechanisms for two-way communication in such regulatory frameworks as the existing Law on Teacher's Status (Law of the Republic of Kazakhstan 2019)—or a new law to be introduced on teachers' associations—would create a better environment for educators to contribute to meaningful dialogue. To wit, Sarinzhipov (2017) suggests that academia should be given definite authority over academic issues and that faculty should have responsibility for the primary areas of teaching and learning, faculty evaluation and status, and the academic part of student life.

Faculty acknowledge that the state plays an important role in providing quality education for all, as education is considered a public good. But for education to be successful, it needs the support of long-lasting partnerships "at all levels of society" (World Education Forum 2000). The "Dakar Framework for Action: Education for All" (World Education Forum 2000, p. 12) and the World Bank (2011) report, "Making Schools Work Accountably", advise more autonomy and accountability for educational institutions, enhancement of administrative and teachers' capacity in school governance, and engagement of various stakeholders at the ground level. The Dakar Framework noted that three factors are required for successful education. These are well-trained teachers, a curriculum that builds upon the knowledge and experience of teachers and learners, and participatory governance and management. This is likewise the case at the level of tertiary education, where the OECD (2018) noted that without more autonomy for HEIs, Kazakhstan will have difficulties in becoming one of the top global economies.

Many countries have been successful in improving their education by providing teaching institutions with more autonomy and accountability, and by adopting a participatory approach to empower various actors in the education systems (World Bank 2011). On the contrary, other countries, with low levels of accountability, are trapped in education systems with poor attainment and high levels of access inequality

(World Bank 2018). Educators, in many developing nations are under-resourced and, hence, have limited analytical capacity to contribute to government-organised meetings (Ratteree 2004).

It should, however, be cautioned that greater autonomy for faculty associations and HEIs will do relatively little by itself for the improvement of national policymaking, because MES absorption of faculty policy inputs depends more on the internal characteristics of the MES than on the characteristics of associations. A community of autonomous educational associations is thus a necessary but not sufficient condition for better policymaking. Teacher and faculty associations contribute to the development of an educational policy community, and in doing so broaden the civil society participation and propose innovation in policy. Though the influence of associations is indirect and slow, it is nonetheless valuable.

Probably, the first thing the MES should do is to reduce faculty workloads; this would enable more research and participation in policymaking. Second, salaries should be raised, so as to attract and retain higher quality talents. Further, faculty should be relieved of MES micro-management and the need to implement and report on marginally useful or unnecessary pedagogical measures. That is, faculty should be trusted to deliver educational quality in their classrooms, as occurs in successful foreign education systems. Intuition suggests that achieving partnerships with employers and outside stakeholders (a topic for separate study) will not be quick to achieve, because these would probably require a more favourable legal environment and greater transparency, which would in turn assure contributors that financial aid and other supports are being used properly. In the meantime, however, to repeat, faculty associations and an educational community could be a catalyst for reforms. Somewhat aside, if the government wishes to improve educational quality quickly, it would be useful to focus efforts on raising a new generation of faculty through pedagogical universities, where teacher pre-service training takes place.

The MES is modernising curricula and is working hard on improving the education workforce at all levels by investing in professional development; educator participation is nonetheless insufficiently recognised in Kazakhstan. Education is inherently a widely held public good, which attracts the interest of students, parents, teachers, faculty, and other stakeholders in the economy and the wider society. As a self-selected professional group which loves learning and its pupils, educators are idealists (Kopeyeva 2019). They are necessarily highly educated, have a strong normative desire for honesty and integrity in higher education

(Rumyantseva and Caboni 2012), and should ordinarily be highly capable of self-organisation and of lobbying for educational interests. There is, hence, good potential for an active and capable education community in Kazakhstan, if given a favourable environment to develop.

Conclusion

The government of Kazakhstan and the MES have demonstrated strong political will to improve education at all levels. This is evident both from the numerous administrative and institutional reform measures introduced, and from rising education budgets. This has resulted in creditable improvements to learning attainment and enrolment rates (inclusion) at primary, secondary, and tertiary levels.

To its credit, the government has evidently recognised that merely throwing money at HEIs would not enhance their performance by as much as desired. Therefore, the government has been wise to extend autonomy, starting with the top universities, prior to raising budgets. The significant corollary here is that, with greater institutional and faculty association capacity set in place, combined with better MES receptivity to faculty initiatives, it could become possible to anticipate more dramatic improvements to educational attainment in the near future.

At the same time, the engagement of outside stakeholders remains low, even though such stakeholders have in some instances come forward. The main impediment is identified by authoritative sources such as the OECD and World Bank, and by many educators, as inflexible MES rules. Therefore, the further encouragement of capacity-building by associations at the secondary and, especially, tertiary levels should become an urgent national task in order to foster innovation in a range of areas of Kazakhstan's education.

Capacity-building has two sides: the faculty association side and the governmental side. On the first side, the evidence indicates that Soviet-era cultural inhibitions in self-organisation are well in the past, for faculty have demonstrated strong determination and ability despite encountering serious difficulties. On the government side, however, bureaucratic rules and structures have a more unyielding character; MES officials have therefore not kept pace with teacher initiatives and their desire for change. Given the high turnover of MES upper and mid-level staff, and consequently the insufficient time available to produce meaningful results, many officials tend to keep the current top-down approach. The noteworthy implication here is that capacity-building on the educational community side, somewhat paradoxically, would be enhanced if there were better partners on the state agency side (see also Soltys 2014).

In sum, both the government and the educational community are moving, even if too slowly for some, in the right direction. Kazakhstan has the potential to develop a high-quality, internationally competitive educational system, if greater organisational and lobbying capacity were to be built within the education community.

As this study of faculty self-organisation is one of the early ones in Kazakhstan, further research should be conducted in order to analyse more closely the interface between faculty associations and education governance.

References

Ahn, E. S., Dixon, J., & Chekmareva, L. (2018). Looking at Kazakhstan's higher education landscape: From transition to transformation between 1920 and 2015. In J. Huisman, A. Smolentseva, & I. Froumin (Eds.), *25 Years of transformations of higher education systems in post-Soviet countries* (pp. 199–127). Palgrave Studies in Global Higher Education. Cham: Palgrave Macmillan. Available via https://link.springer.com/chapter/10.1007/978-3-319-52980-6_8. Accessed 22 March 2020.

Anderson, K. (2006). Education in Central Asia during the transition to a market economy. In S. P. Heyneman & A. J. De Young (Eds.), *The challenges of education in Central Asia* (pp. 131–152). Charlotte, NA: Information Age Publishing.

Asanova, J. (2007). Teaching the canon? Nation-building and post-Soviet Kazakhstan's literature textbooks. *Compare, 37*(3), 325–343.

Bascia, N. (Ed.). (2015). *Teacher unions in public education: Politics, history and the future*. New York: Palgrave Macmillan.

Bridges, D., & Sagintayeva, A. (2014). Introduction. In D. Bridges (Ed.), *Educational reform and internationalisation: The case of school reform in Kazakhstan* (pp. xxii–xlii). Cambridge: Cambridge University Press.

Carr, J., & Beckett, L. (2018). *Teachers and teacher unions in a globalised world: History, theory and policy in Ireland*. London: Routledge.

De Witt, N. (1961). *Education and professional employment in the USSR*. Washington: National Science Foundation.

Emrich-Bakenova, S. (2009). Trajectory of civil service development in Kazakhstan: Nexus of politics and administration. *Governance, 22*(4), 717–745.

Emrich-Bakenova, S. (2019). Civil service development in Kazakhstan. In A. Farmazmand (Ed.), *Global encyclopedia of public administration, public policy, and governance*. New York: Springer.

European Commission. (2015). *ECTS users' guide*. Available via https://ec.europa.eu/education/ects/users-guide/docs/ects-users-guide_en.pdf. Accessed 22 March 2020.

European Commission. (2017). *Overview of the higher education system—Kazakhstan*. Brussels: European Commission.

Expert.gov. (2018). *Kazakhstan country commercial guide: Kazakhstan—Education*. September 19. Available via https://www.expert.gov/article?id=Kazakhstan-Education. Accessed 25 January 2019.

Expert.gov. (2019). *Kazakhstan country commercial guide: Kazakhstan—Education*. Available via https://www.expert.gov/apex/article2?id=Kazakhstan-Education. Accessed 22 March 2020.

Fimyar, O. (2014). "Soviet", "Kazakh" and "World Class" in the contemporary construction of educational understanding and practice in Kazakhstan. In D. Bridges (Ed.), *Education reform and internationalisation: The case of school reform in Kazakhstan* (pp. 177–195). The Cambridge Education Research Series. Cambridge: Cambridge University Press.

Fimyar, O., Yakavets, N., & Bridges, D. (2014). Educational reform in Kazakhstan—The contemporary policy agenda. In D. Bridges (Ed.), *Education reform and internationalisation: The case of school reform in Kazakhstan* (pp. 53–68). The Cambridge Education Research Series. Cambridge: Cambridge University Press.

Frost, D., Fimyar, O., Yakavets, N., & Bilyalov, D (2014). The role of the school director in educational reform in Kazakhstan. In D. Bridges (Ed.), *Education reform and internationalisation: The case of school reform in Kazakhstan* (pp. 217–238). The Cambridge Education Research series. Cambridge: Cambridge University Press.

Gappa, J. M., Austin, A. E., & Trice, A. G. (2007). *Rethinking faculty work: Higher education's strategic imperative*. San Francisco: Jossey-Bass.

Heyneman, S. P. (2008). Three universities in Georgia, Kazakhstan and Kyrgyzstan: The struggle against corruption and for social inclusion. *Prospects, 37*, 305–318. UNESCO IBE. Available via http://ten.edc.org/docs/3.pdf. Accessed 23 March 2020.

Information Analytic Centre (IAC). (2018). *National report on the current state and development of the education system in the republic of Kazakhstan for the year 2017*. Nur-Sultan: Publishing service of the IAC.

IQAA. (2016). *Higher education and quality assurance of higher education in the republic of Kazakhstan*. Nur-Sultan: Independent Kazakh Agency for Quality Assurance in Education.

Katsu, S. (2017). Higher education: Medium and long-term challenges. In M. Hartley & A. Ruby (Eds.), *Higher education reform: The case of Kazakhstan* (pp. 179–190). Cambridge: Cambridge University Press.

Kazakhstan Association of Young Researchers and Faculty (KAYRF). (2018, April 26). Statement of Kazakhstan Association of Young Researchers and Faculty Meeting. Kazakh National Women's University, Almaty [in Russian]. Available via http://kazmkpu.kz/wp-content/uploads/2018/05/%D0%A0%D0%95%D0%97%D0%9E%D0%9B%D0%AE%D0%A6%D0%98%D0%AF-%D0%BD%D0%B0-%D1%80%D1%83%D1%81.pdf. Accessed 27 March 2020.

KIMEP University. (2019). Department of Quality Assurance and Institutional Research, internal document, Almaty.

Kopeyeva, A. (2019, April 17). Kazakhstan's teachers: Underpaid, overburdened, and undervalued. *The Diplomat*. Available via https://thediplomat.com/2019/04/kazakhstans-teachers-underpaid-overburdened-and-undervalued/. Accessed 23 March 2020.

Law of the Republic of Kazakhstan. (2001). *On Non-commercial Organisations* [in Russian]. Law No 142 of 16 January 2001.

Law of the Republic of Kazakhstan. (2007). *On Education* [in Russian]. Law No. 319-III of 27 July 2007.

Law of the Republic of Kazakhstan. (2019). *On Teachers' Status* [in Russian]. Law No. 293-VI of 27 December 2019.

Massyrova, R., Tuatenbaeva, A., Tussupova, A., Zhalova, A., & Bissenbayeva, Z. (2015). *Changes in the higher education system of Kazakhstan*. Paper for the 3rd World Conference on Psychology and Sociology, 6–8 November 2014, Izmir, Turkey. *Procedia-Social and Behavioral Sciences, 185,* 49–53. Available via https://www.sciencedirect.com/science/article/pii/S187704281502251X. Accessed 19 May 2020.

Mehisto, P., Kambatyrova, A., & Nurseitova, K. (2014). Three in one? Trilingualism in educational policy and practice. In D. Bridges (Ed.), *Education reform and internationalisation: The case of school reform in Kazakhstan* (pp. 152–176). The Cambridge Education Research Series. Cambridge: Cambridge University Press.

Nazarbayev, N. A. (2006, May 26). *To the knowledge economy through innovations and education* [in Kazakh]. Nur-Sultan: Lecture of the President of the Republic of Kazakhstan N.A. Nazarbayev at the Eurasian National University of L.N. Gumilev.

Nusche, D., Radinger, T., Falch, T., & Shaw, B. (2016). *Review of school resources: Denmark 2016*. Paris: OECD. Available via https://books.google.kz/books?id=nWSQDwAAQBAJ&pg=PA100&dq=teacher+unions+international+evidence&hl=en&sa=X&ved=0ahUKEwjCs_-YnPvkAhV8xcQBHdS-D48Q6AEIJzAA#v=onepage&q=teacher%20unions%20international%20evidence&f=false. Accessed 23 March 2020.

OECD. (2007). *Higher education in Kazakhstan 2007*. Reviews of National Policies for Education. Paris: OECD. Available via https://doi.org/10.1787/9789264268531-en. Accessed 23 March 2020.

OECD. (2018). *Education policy outlook—Kazakhstan*. Paris: OECD. Available via http://www.oecd.org/education/Education-Policy-Outlook-Country-Profile-Kazakhstan-2018.pdf. Accessed 23 March 2020.

OECD/World Bank. (2015). *OECD reviews of school resources: Kazakhstan 2015*. Paris: OECD. Available via https://doi.org/10.1787/9789264245891-en. Accessed 23 March 2020.

Plessius, H., & Ravesteyn, P. (2010). The paradox of more flexibility in education: Better control of educational activities as a prerequisite for more flexibility. In N. Reynolds & M. Turcsányi-Szabó (Eds.), *Key competencies in the knowledge society* (Vol. 324, pp. 301–309). KCKS 2010. IFIP Advances in Information and Communication Technology. New York: Springer.

Quacquarelli-Symonds (QS). (2019). *World university rankings*. Available via https://www.topuniversities.com/university-rankings/world-university-rankings/2019. Accessed 23 March 2020.

Ratteree, B. (2004). *Teachers, their unions and the education for all campaign*. Paper commissioned for the EFA Global Monitoring Report 2005, The Quality Imperative. Paris: UNESCO. Available via https://unesdoc.unesco.org/ark:/48223/pf0000146693. Accessed 27 March 2020.

Raza, R. (2009). *Examining autonomy and accountability in public and private tertiary institutions*. Washington: World Bank Human Development Network.

Rouse, M., Yakavets, N., & Kulakhmetova, A. (2014). Towards inclusive education: Swimming against the tide of educational reform. In D. Bridges (Ed.), *Education reform and internationalisation: The case of school reform in Kazakhstan* (pp. 196–216). The Cambridge Education Research Series. Cambridge: Cambridge University Press.

Ruby, A., & McLaughlin, C. (2014). Transferability and the Nazarbayev Intellectual Schools: Exploring models of practice transfer. In D. Bridges (Ed.), *Education reform and internationalisation: The case of school reform in Kazakhstan* (pp. 287–300). The Cambridge Education Research Series. Cambridge: Cambridge University Press.

Rumyantseva, N. L., & Caboni, T. C. (2012). Are academics in Kazakhstan capable of self-regulation? A study of faculty's normative structure in the midst of higher education decentralisation reforms. *Tertiary Education and Management, 18*(1), 1–16.

Sagintayeva, A., Hartley, M., Zhakypova, F., & Apergenova, R. (2017). The road to autonomy: Governance reforms in Kazakhstan's system of higher education since independence. In M. Hartley & A. Ruby (Eds.), *Higher education reform: The case of Kazakhstan* (pp. 13–28). Cambridge: Cambridge University Press.

Sarinzhipov, A. (2017). The imperative of faculty participation in university governance for higher education development in Kazakhstan. In M. Hartley & A. Ruby (Eds.), *Higher education reform: The case of Kazakhstan* (pp. 157–175). Cambridge: Cambridge University Press.

Satubaldina, A. (2018, October 6). Kazakhstan to allocate nearly $763 million to build 52 schools. *The Astana Times*. Available via https://astanatimes.com/2018/10/kazakhstan-to-allocate-nearly-763-million-to-build-52-schools/. Accessed 23 March 2020.

Seedstars.com. (2020, January 6). *Born to code: How Kazakhstan is growing the future generation of tech talents*. Seedstars. Available via https://www.seedstars.com/magazine/born-code-how-kazakhstan-growing-future-generation-tech-talents/. Accessed 23 March 2020.

Sen, A. (2000). *Development as freedom*. New York: Anchor Books.

Shamshidinova, K., Ayubayeva, N., & Bridges, D. (2014). Implementing radical change: Nazarbayev Intellectual Schools as agents of change. In D. Bridges (Ed.), *Education reform and internationalisation: The case of school reform in Kazakhstan* (pp. 71–82). The Cambridge Education Research Series. Cambridge: Cambridge University Press.

Soltys, D. (1997). *Education for decline: Soviet vocational and technical schooling from Khrushchev to Gorbachev*. Toronto: University of Toronto Press.

Soltys, D. (2014). Challenges to institutionalisation of environmental NGOs in Kazakhstan's corporatist policy arena. *Journal of Contemporary Asia, 44*(2), 342–362.

Soltys, D. (2015). Similarities, divergence, and incapacity in the Bologna Process reform implementation by the former-socialist countries: The self-defeat of state regulations. *Comparative Education, 51*(2), 179–195.

Soltys, D., & Orynbassarova, D. (2013). Delivering environmental education in Kazakhstan through civic action: Second-wave values and governmental responses. *Environmental Values, 22*(1), 101–122.

Tempus. (2010). *Higher education in Kazakhstan*. Available via http://tempuskaz.belight.net/downloads/kazakhstan_review_of_higher_education. Accessed 26 May 2020.

TheGlobalEconomy.com. (n.d.). *Kazakhstan: Education spending, percent of GDP*. UNESCO sources. Available via https://www.theglobaleconomy.com/kazakhstan/education_spending. Accessed 23 March 2020.

Turner, F., Wilson, E., Ispussinova, S., Kassymbekov, Y., Sharimova, A., Balgynbayeva, B., & Brownhill, S. (2014). Centers of excellence. In D. Bridges (Ed.), *Education reform and internationalisation: The case of school reform in Kazakhstan* (pp. 83–105). The Cambridge Education Research Series. Cambridge: Cambridge University Press.

Van Keuren, H. (2011). Influencing the status of teaching in Central Asia. In I. Silova (Ed.), *Globalisation on the margins: Education and post-socialist transformations in central Asia* (pp. 173–201). Charlotte, NC: Information Age Publishing.

Wilmoth, G. D. (2011, November 7–18). *Central Asia's future role in international higher education*. Paper presented at the international conference on contemporary trends in education. Nazarbayev University, Nur-Sultan, Kazakhstan. Available via https://wilmoth.com.au/publications/WilmothNazUnivPaper.pdf. Accessed 23 March 2020.

Wilson, E. (2017). Alternative paths to upgrading existing teacher qualifications: The Kazakhstan-based Centre of Excellence Teacher Education Programme. In M. Hartley & A. Ruby (Eds.), *Higher education reform: The case of Kazakhstan* (pp. 107–124). Cambridge: Cambridge University Press.

Winter, L., Rimini, C., Soltanbekova, A., & Tynbayeva, M. (2014). *The culture and practice of assessment in Kazakhstan*. In D. Bridges (Ed.), *Education reform and internationalisation: The case of school reform in Kazakhstan* (pp. 133–151). The Cambridge Education Research Series. Cambridge: Cambridge University Press.

World Bank. (2011). *Making schools work—New evidence of accountability reforms* (B. Bruns, D. Filmer, & H. A. Patrinos, Eds.). Washington, DC: World Bank Group. Available via http://siteresources.worldbank.org/EDUCATION/Resources/278200-1298568319076/makingschoolswork.pdf. Accessed 23 March 2020.

World Bank. (2017). *Kazakhstan—Education modernisation project*. Washington, DC: World Bank Group. Available via http://documents.worldbank.org/curated/en/340301488682834835/pdf/Kazakhstan-PAD-02092017.pdf. Accessed 23 March 2020.

World Bank. (2018). *World development report 2018: Learning to realise education's promise*. Washington, DC: World Bank Group. Available via https://www.worldbank.org/en/publication/wdr2018. Accessed 23 March 2020.

World Education Forum. (2000, April 26–28). *Dakar framework for action. Education for all: Meeting our collective commitments*. Dakar, Senegal: World Education Forum. Available via http://www.undp.org.lb/programme/governance/institutionbuilding/basiceducation/docs/dakar.pdf. Accessed 23 March 2020.

Yakavets, N. (2014). Educational reform in Kazakhstan: The first decade of independence. In D. Bridges (Ed.), *Education reform and internationalisation: The case of school reform in Kazakhstan* (pp. 1–26). The Cambridge Education Research Series. Cambridge: Cambridge University Press.

Yakavets, N., & Dzhadrina, M. (2014). Educational reform in Kazakhstan: Entering the world arena. In D. Bridges (Ed.), *Education reform and internationalisation: The case of school reform in Kazakhstan* (pp. 28–52). The

Cambridge Education Research Series. Cambridge: Cambridge University Press.

Yakavets, N., Frost, D., & Khoroshash, A. (2015). School leadership and capacity building in Kazakhstan. *International Journal of Leadership in Education, 20*(3), 345–370.

Yedgenov, B (2019, September 3). *Impact evaluation: Overview, randomised control trial*. Paper presented at the impact evaluation workshop, Almaty Management University and Georgia State University, Almaty, Kazakhstan.

Young Researchers Alliance. (2018). *Charter of Non-Governmental Organization Young Researchers Alliance* [in Russian]. Available via http://www.yrascience.kz/Ustav-OOO-YRA_rus.pdf. Accessed 27 March 2020.

PART II

New Paradigms for Kazakhstan's Market Economy

CHAPTER 5

Public Service Provision in Kazakhstan: Challenges to Procedural Fairness

Saule Emrich-Bakenova

Introduction

Since securing national independence in 1991, Kazakhstan, like other former Soviet republics, has undergone major political, economic, and social transformation. The changes have included reforms of constitutional order, state organisation and management systems, and economic and legal structures (Emrich-Bakenova 2019).

Kazakhstan's achievements, as expressed in international performance indicators, are uneven. For example, relative to other countries in Central Asia, Kazakhstan is praised by international observers for successful economic development and dynamism, professionalisation of public administration, and adherence to the rule of law (Deppe and Pudelka 2014; Organisation for Economic Cooperation Development (OECD) 2017; Institute for Management Development (IMD) 2019). In particular, Kazakhstan's position in the World Competitiveness Ranking has

S. Emrich-Bakenova (✉)
Department of Public Administration and International Development,
KIMEP University, Almaty, Kazakhstan
e-mail: emrich.bakenova@kimep.kz

A. Koulouri and N. Mouraviev (eds.),
Kazakhstan's Developmental Journey,
https://doi.org/10.1007/978-981-15-6899-2_5

greatly improved from being 47th in 2016 to 32nd in 2017, and 34th in 2019 (IMD 2019; Suleimenova et al. 2019). Yet Kazakhstan's key attractiveness indicators are ranked low in competency of government (15%) and effective legal environment (13.6%) and high in business-friendly environment (61.4%), and policy stability and predictability (54.3%). According to the World Bank, Kazakhstan's government effectiveness (including such factors as corruption, quality of regulation, accountability, political stability, rights, and civil liberties) has grown by 25% over the last five years, to a rank of 86th with an index value of 0.01 (where −2.5 is weak and 2.5 strong). Meanwhile, according to the rule of law index, Kazakhstan occupies the 117th position with an index value of −0.41 (World Bank 2017; Suleimenova et al. 2019).

These various indexes and rankings show that one should be cautious regarding making generalisations about the country's development. Although there are consistent achievements in one area, there are quite significant consistent failures in other areas. In this chapter, the discussion focuses on exactly this type of contradiction exemplified by the country's achievements in the development of market-oriented and technology-driven public services, and paradoxical underdevelopment, even stagnation, when it comes to administrative procedural law.

Drawing on the analysis of certain cases related to disputes between citizens and administrative bodies, and on the review of laws, government reports, policy documents, and analytical expert positions, this chapter argues that the focus on automation of government decisions, on the use of technological tools, and on the treatment of citizens[1] primarily as consumers of public services effectively overshadows the procedural side of administrative actions, which is viewed as circumstantial. The focus of public administration activities on service provision "controlled by customer satisfaction" (Brewer 2007, p. 553) and on the final results negates the fact that just and fair administrative procedures are also "part of the outcome to be expected from good government" (Gleeson 2006, p. 11). Public administrators are expected to act not only within the parameters of managerial benchmarks but also to be guided in their actions by the principles of administrative justice.

[1] The term "citizens" is a shorthand for all the categories of persons (including legal entities) who are affected by the decisions of public bodies.

The first part of the chapter provides a review of recent public service reforms in Kazakhstan with specific focus on the relationship between service provision and protection of citizens' rights and interests in the decisions made by public officials. Further, the chapter reviews the state of procedural law in Kazakhstan. Finally, potential implications are identified with respect to the value that procedures and principles of administrative justice have in Kazakhstan.

Public Service Reform

Concerns regarding the country's economic development focus on providing a favourable investment climate and on building investor confidence led Kazakhstan to initiate public administration reforms within the framework of the fashionable model of the service-oriented state, which was inspired by corporate governance and new public management (NPM) (Knox 2008; Bhuiyan and Amagoh 2011; Janenova and Knox 2017). This came about through the introduction of standards for public services and procedures (public service charters) for all government organisations, as well as the establishment of one-stop shops and an e-government system (Janenova and Kim 2016; OECD 2017). Within the framework of corporate management, in 2013, Kazakhstan adopted rather briskly the Law on Public Services and established the state corporation "Government for Citizens" through the merger of several state enterprises. This integration, as government envisaged, was supposed to increase effectiveness and cost efficiency of public administration bodies under the slogan of "better quality for less money" by eliminating contradictions in documents (rules, standards, regulations) and reducing administrative corruption (Janenova and Kim 2016; Janenova and Yesdauletov 2017; Kazakhstan 2018; Parliament.kz 2019).

In 2012, a large number of deficiencies in public service provision were noted, such as "450 violations just for one type of provided public service, red tape resulted in 500 documents per one official, and citizens were required by public administration bodies to provide more than 22 million documents just in one year" (Nazarbayev 2013). Since then Kazakhstan's legislators have introduced changes to 40 laws and 10 codes with a total of 578 amendments that aimed to reduce the required documents for public service users by 100 million within a year (Parliament.kz 2019).

Highlighting values such as professionalism and patriotism, the civil service in Kazakhstan is also orientated towards performance results and innovative approaches that would foster them. According to the government, the fast-paced culture of civil service does not support the tradition of controlling the process, which generates red tape and lowers the efficiency (CSA-AC 2018). Public agencies at the national and regional levels are now assessed based on the effectiveness of their interactions with citizens in the provision of public services, consideration of citizens' complaints, and transparency and openness (Kazakhstan 2018). A recent official infographic report boasts improved performance results, as exemplified in the decrease in citizens' complaints to and about public administration bodies by approximately 14 times (from 23 million in 2007 to 1.7 million in 2017). Moreover, there are overall fewer failures to conform to the standards of service provision, for example, the demands that citizens should provide additional (deemed unnecessary) documents have decreased and the time for consideration of citizens' complaints and requests have shortened (Kazakhstan 2018).

Obsessed with technologies designed to process standard services, public officials in many public administration systems, including Kazakhstan, now use information and communication technologies to perform functions that previously would have been carried out by employees. The e-government system not only allows citizens of Kazakhstan to track the status of their messages, electronic applications, queries, and official letters sent in a digital form to various government agencies, but also to receive, within a defined period of time, government agencies' responses (Kassen 2017, 2018). The government projects that by 2020, 90% of public services will be delivered in electronic form[2] (Kazakhstan 2018; Parliament.kz 2019). The universal digital platform also serves as a managerial instrument to continuously assess performance of civil servants using achieved results for their career progression (Kassen 2018).

In addition, civil servants also maintain personal blogs, that is, electronic platforms where they cover a range of topics including public services, health care, and education. These blogs aim to ensure direct dialogue between the public and top government officials (e.g. ministers, local government heads) (Kassen 2018). However, as reviewers emphasise, information on these platforms is frequently too generic, reminiscent of "window dressing" for a "new phenomenon … often called electronic formalism" of e-bureaucracy (Kassen 2018, p. 10; see also Tokayev 2020).

[2] In 2018, 446 out of 746 public services were delivered online via the e-government portal (Kazakhstan 2018).

Thus, in a relatively short period of intensive implementation and with strong political support, the public administration of Kazakhstan achieved, as claimed, provision of more transparent information about public services, increased accessibility of public services both through face-to-face interaction and e-government application, and higher standards of professionalism of public service staff (Janenova and Kim 2016; Janenova and Yesdauletov 2017).

Public Service Provision: Consumers Versus Citizens

Promoting the image of public administration as a corporation, or else, as a "supermarket delivering a wide variety of public services" (Olsen 2005, p. 6) puts emphasis on revenue generation and commercialisation of public administration functions (Brewer 2007). Yet the ideology of consumerism is supposed to be balanced on a daily basis with routine application of administrative rules in issuing licences, permits, certificates, social benefits, etc. These activities require from public officials not only competence in the knowledge of laws but also a level of professionalism to make such decisions that take into consideration, in a proportionate way, interests of the state along with the legitimate rights and interests of individual citizens. The choices the public workers make can ultimately have a tremendous impact on the lives of the general public (Bovens and Zouridis 2002).

Public officials by way of their action or inaction (e.g. alterating their initial decision or keeping administrative silence on citizens' appeals) are "significant and salutary drivers of constitutional implementation" (MacDonnell 2015, p. 386). Almost any constitutional norm, especially those that regulate rights and freedoms of citizens, presumes existence of administrative norms that one way or another promote or limit constitutional guarantees (Podoprigora 2010a). Therefore, constitutional norms remain purely symbolic when the mechanism of their implementation, the procedural side, is absent or disregarded. Specifically, in this area conflicts arise more often between citizens and public officials (Podoprigora, n.d.). To clarify the argument, the following three subsections provide certain examples of situations when the procedural side of the service provision was largely disregarded.

Outsourcing Public Services

A variety of labour-intensive public services, such as health care, social care, education, and training, have been progressively delivered by quasi-public agencies (joint-stock companies, limited liability companies, state institutions) and those representing the business and voluntary sectors. While implementation of NPM by the public sector agencies coexists, to a large extent, with requirements of legalism in decisions, with the transfer of public services to the new agents of the state, the decisions are made according to "market culture and values in the first place" with the focus on speedy access to or low cost of the service, as opposed to due process and fairness in treatment (Benish 2014, p. 117). This undoubtedly raises concerns regarding administrative justice because of the weakening of "administrative and legal values" and not least because "delegalised and debureaucratised" agents are not subject to the public sector requirements and administrative norms. They enjoy "much more discretion and flexibility in their day-to-day operation" (Benish 2014, p. 117). For example, in Kazakhstan, the Law on Administrative Procedures (which is discussed in the section on the Law on Administrative Procedures: the critical legal framework) only regulates public administration bodies and does not cover other organisations that are granted public functions and right to issue administrative acts. Meanwhile, activities of these organisations, as evaluation shows, frequently have questionable and uncertain outcomes, lack transparency, and raise significant concerns regarding proper governance (OECD 2017).

Automated Decisions

Several studies of public management (service) reforms explicitly take into account the outcome of procedures. A majority of the studies consider administrative procedures as "complex and cumbersome" and essentially constituting a "rule burden" (or red tape), which is a source of inefficiency and ineffectiveness (Kaufmann and Feeney 2014, pp. 179, 181). Meanwhile, the 'argument that red tape is not about rules and procedures per se, but has more to do with the "relative (in)effectiveness of rules" is buried in the discussion of public service reforms (Kaufmann and Feeney 2014, p. 181) and subdues the issue of what might happen outside of standard situations when technologies fail. On the one hand, services that are delivered via web-based platforms, without human intervention, may leave hardly any margin "for the arbitrary exercise of power in implementing rules and regulations" (Bovens and Zouridis 2002, pp. 180–181). New technologies make, or at least have the capacity to make, public services more efficient, interactive, and accessible. However, automated programmes are made in such a way that any deviation from standard practice may immediately turn into an unresolvable problem with blame being put on the automated programme itself (Podoprigora 2015). While transparent and understandable to a technician, an automated system could be "opaque" to those citizens who are affected by a particular decision for which this system was used, for example a calculation of social assistance benefits or assessment of taxes or fines (Zalnieriute et al. 2019, p. 442). There is high risk of mistakes or premature decisions on the part of public officials. Blind application of the law may lead to arbitrariness, exactly "because no account is taken of the circumstances of the case" (Bovens and Zouridis 2002, p. 182). Not only might the citizen's problem appear to be outside of the automated algorithm programmed by the IT expert, but the decision itself would then be lacking transparency without clear understanding of the variables, inputs and outputs, used to arrive at it (Zalnieriute et al. 2019).

Difference in Complaints (Appeals)

Snapshots of performance results may create a sense of administrative bodies' accountability. They also could be an appealing way of presenting complex activities and policies to the citizens. For example, Kazakhstan's government identifies a decrease in the number of citizens' complaints in the last several years and indicates that there is potential for a further decrease in volume because only "on 46 percent of requests there were given clarifications". In other words, public administration bodies need to reach a further 54% increase in production of clarifications to citizens regarding their complaints (Kazakhstan 2018). The performance results infographics inadvertently show an optimistic, non-litigious relationship between citizens and public administration.

Another example is the Open Almaty website open-almaty.kz (established in the largest city of Kazakhstan, Almaty), which is presented as an interactive guidebook that contains information on over 700 life situations and is aimed at guiding citizens in their quest to resolve their own problems (e.g. resolution of issues with municipal transport, matters of public education or social assistance, or payment of taxes and fines, or simply access to general official information). The interactive, user-friendly portal boasts handling over 154,000 complaints, appeals, and information requests made by the residents of the mega-city between 2017 and 2019, with over 70% of the issues being resolved on the spot (Bizhikeyeva 2019).

Government indicates concern to improve services and reduce their costs by encouraging fast and easy processing of various complaints, appeals, queries, and applications (including online). Undoubtedly, fast responses tend to receive positive feedback from users of public services. However, to what extent can quantified and measured performance outcomes provide sufficient information regarding how they were achieved and whether consideration of citizens' claims were handled fairly?

Opponents of NPM argue that performance results highlight a too narrowly defined relationship that public service consumers have with the government that is "focused on personal consumer satisfaction with received services" (Brewer 2007, p. 553). Not least customer-oriented systems obscure what happens when citizens have broader concerns related to "public sector administrative mechanisms through which" citizens seek redress (Brewer 2007, p. 550). It is therefore critical to differentiate in this way between certain types of complaints and appeals. There are those complaints (appeals) that may arise in the course of general activities of public administration and that may concern general unspecified categories of citizens. Citizens may focus their complaints on, for example, lack of information related to the services or some technical difficulties in receiving the registration, or appeals about condition of roads, schools, or hospitals. The other category of complaints, however, may concern administrative decisions (acts) directed to an individual or a specific group of citizens. Citizens may be upset in cases related to land (e.g. seizure of private land for state needs or the right to privatise a piece of land), or in cases related to the right to social housing, or as a result of ubiquitous migration issues (e.g. annulment of permanent residency, request for refugee status, or application for citizenship, or deportation), or in cases related to the right to hold a public meeting. They may disagree with an administrative decision either in substance or in procedure (or both). Disagreements regarding a procedure lead to the disputes that draw attention to the procedural side of administrative action (or inaction). Citizens may dispute the administrative decisions (acts) and in due course they may encounter administrative silence, or they may find that the administrator failed to provide sufficient reasons for the decision, or they may encounter bias in treatment, or perhaps they were not heard before the final decision was made, or they find that interests of the state disproportionally outweigh their individual interests.

These are only a few examples that highlight the difference between public services conceived narrowly in a consumerist market mode of interaction between a citizen and a public official and those interactions that necessarily focus on enforcement of citizens' individual rights in a constitutional state. In other words, it is crucial to differentiate the role of the public administration as a "seller", or provider, of services from a much broader role: that of a guarantor of citizens' rights in the process of receiving those services.

As Kassen (2018, p. 13) argues, "it would be naïve to presume that new collaborative digital platforms alone could quickly change the bureaucratic nature of many public institutions and further digital democracy in a real-life context" since "the historical oppressive past of [Kazakhstan's] society, … [its] controversial political context continues to have a crucial impact on all public sector reforms" including those of the administrative justice system. Despite obvious advantages of e-government as a form of relationship between the state and a citizen of Kazakhstan, citizens are in fact more deprived (compared to traditional public services provided live in the public offices) of the possibility to present their explanations and participate in procedures that affect their interests (Podoprigora 2014, 2012a).

Administrative Procedures in Kazakhstan

Administrative Complaints: Results

As mentioned in the section on Public service reform, from 2007 to 2017, the overall number of citizens' complaints to public officials in Kazakhstan decreased drastically. On the contrary, the number of grievances that were brought to administrative bodies against their decisions (actions and inactions) increased as a percentage of the total number of complaints (Table 5.1). This trend demonstrates the ongoing presence of conflict between citizens and public administration bodies, representing an important change for a post-Soviet country, where formerly it was taken for

Table 5.1 Complaints and administrative appeals in Kazakhstan, 2012–2017

Year	*All complaints and appeals*	*All complaints*	*Administrative appeals*	*Repeated administrative appeals when a citizen was not satisfied with first decision*
2012	19.1 million	126,866 (0.7%)	10,746 (8.5%)	5534 (0.03%)
2013	10.8 million	127,020 (1.37%)	n/d	2749 (0.03%)
2016 (10 months)	2.1 million	155,156 (7.3%)	21,586 (13.9%)	1246 (0.07%)
2017	1.7 million	26,600 (1.5%)	6132 (23%)	n/d

Source Compiled by the author from prokuror.gov.kz; Kazakhstan (2018)

granted that there could not be conflicts between citizens and the state (Podopriogora 2015).

The low number of second-time submission of appeals (Table 5.1) may be considered a positive indicator, assuming that most matters were dealt with in a first review. However, this could also be an indicator that complaints were not dealt with satisfactorily and citizens opted for judicial review, or citizens experienced administrative appeal fatigue that prompted them to withdraw from the appeal process entirely.

The presented data should, in general, be viewed with caution. This is because there are multiple formally acknowledged instances when public administration bodies have failed to consider citizens' complaints; or have failed to indicate information about a decision made on the basis of appeal; or have made incorrect and incomplete entries of information related to citizens' complaints, including incorrect time of complaint registration which then may lead to viewing a complaint as filed after the deadline and therefore not requiring consideration. This in turn leads to incorrect accounting and statistics. In 2012, for example, more than 43,000 facts of incorrect and incomplete data entries related to appeals were identified (Prokuror Kostanai 2015). Given this context, the number of citizens' complaints could be even higher than indicated in Table 5.1. For example, the 2018 statistics by the General Prosecutor present different numbers and show the increasing trend of complaints received by public administration bodies since 2015. Specifically, in 2015, public administration considered 7068 complaints; in 2016, 25,115; in 2017, 38,824; and for the period of the first nine months in 2018, 33,177 complaints (IZ 2018).

The Law on Administrative Procedures: The Critical Legal Framework

Administrative appeals in Kazakhstan are regulated by the Law on Administrative Procedures 2000 (LAP). This law is supposed to provide an overarching procedural framework for various administrative activities of all public administration bodies. There is also the Law on Considerations of Complaints Brought by Persons and Legal Entities 2007 (LCC), which covers the procedural side of the consideration of citizens' complaints, appeals, and requests. However, a number of provisions in these two legal documents contain substantial redundancies, which in itself is a problem of the LAP. This chapter focuses on the LAP.

Among many provisions, the LAP states that the responsibility of public officials is to make lawful and reasoned decisions. In cases of a citizen's appeal of a decision, public officials are required to make a comprehensive review of the complaint (appeal) and to produce a reasoned written response.

In their acts, administrative officials are also supposed to adhere to the LAP's principles. The principles have binding legal force in administrative practice, which aims to guarantee stability and predictability of actions of public administration bodies in relation to citizens (Ametistova and Gabbasov 2010; Porokhov 2010). There are 11 principles, among them the rule of law; equality before the law; mutual responsibility and balance of interests of individuals, society, and state; precedence of rights and freedoms of citizens; and prevention of bureaucratisation.

Thus, formally public officials in Kazakhstan are guided in their actions and decisions by procedural provisions and principles as per the LAP. The LAP, in its principle 9, provides a basis for other laws, rules, and standards regulating various activities of public administration bodies at the national and subnational levels. Article 24 of the law proclaims that its provisions aim to be "foundational", that is to provide a framework for the development and adoption of normative legal acts that regulate other types of administrative procedures. The objective of the LAP is also noble. The legislators outline that it aims to establish administrative procedures to foster improvement of public administration activities, provide uninterrupted functioning of public administration bodies, and encourage prompt administrative decision-making focused on citizens' rights and freedoms (LAP preamble).

Notwithstanding the formally stated goals, the LAP, however, does not deliver what it promises. The LAP failed to be a law of the "first" level for other normative acts (Gabbasov 2014). Not only does the LAP fail to be a systemic, core framework law for all administrative procedures, but it also lacks direct effect, and principally contains a set of declarations and overabundance of blanket norms (Podoprigora 2015; Gabbasov 2014, 2015). To some critics, the LAP is "so empty, that its application or non-application is hardly noticeable" (Deppe and Pudelka 2014, p. 1).

The law also contains a list of lengthy exclusions (i.e. contexts in which it does not apply), including consideration of citizens appeals if those are reviewed within the code of administrative offences, or criminal and civil procedural codes, or any other acts the development of which is regulated by laws other than the LAP. It appears that administrative procedures described by the LAP are supposed to be followed in the activities of public administration bodies only if those bodies are not regulated by other normative legal acts. In other words, the LAP is used to fill a legal gap (Gabbasov 2014).

The Relationship Between Administrative Procedures and Public Services

A strong focus on building the image of the service-oriented state raises the issue of the relationship between administrative procedures and public services as Podoprigora observed (2013, 2015). He asks, are the administrative procedures a form in which public services are provided or are public services a type of administrative procedure? However, these questions did not receive serious examination and discussion in Kazakhstan. Prior to 2013 (when the Law on Public Services was adopted) public services were regulated as part of the LAP. Adoption of a separate Law on Public Services made it easy then to neglect the crucial point that public services are only *external, standardised* forms of administrative procedures. To deliver these services, administrators adopt further administrative acts (Podoprigora 2013). Experts argue that administrative procedures are broader than public services and therefore the LAP should be viewed as a framework law for all types of activities, such as education or housing, and also for special laws, such as the Law on Public Services.

Missing Procedural Oversight

Fragmentation (adoption of a special law on public services), lack of a systemic and holistic approach to administrative procedures (e.g. provision of public education follows its own procedures, as do taxation and immigration, etc.) reinforces another issue. Administrative appeal as such lacks a proper system of control. Separate and therefore lacking cohesion, existing regulations do not determine responsibility of public administration bodies and their officials for violations of the procedure and terms of consideration of citizens' complaints neither by the LAP nor by the LCC and legal entities. An exception is Article 189 of the Administrative Code of Offences but even this provision affects only entities belonging to large organisations (IZ 2018).

Lack of procedural oversight, however, serves to protect interests of public administration bodies. These bodies have advantage in disputes with citizens as statistics and analytics demonstrate. The largest number of disputes arise in the spheres of taxation, employment and pension, land, and internal affairs (see Shatov 2012; Shormanbayeva 2012; Podoprigora 2012b, 2013). For example, administrative consideration of complaints related to taxation disputes shows that, in 2018, two out of four complaints made by citizens were satisfied and, out of 103 complaints made by legal entities, only 23 were satisfied and 11 were satisfied only partially, with the rest being left as they were. While there are those complaints that are not substantiated, there are also those that require consideration. Experts warn that citizens' complaints frequently fall victim of underdeveloped administrative review process and unfairness (IZ 2018; Policy 2019). Higher ranking public administration bodies or officials as a rule do not satisfy the complaint of a citizen even if the decision was incorrect or outright unlawful. This is an obvious shortcoming of not only the procedural law but of an existing legal culture of public administration in Kazakhstan (Policy 2019). Bureaucrats are afraid of making decisions especially when the law does not provide a detailed algorithm for actions, instead granting administrators discretion (Kapital.kz 2017; Tokayev 2020). In turn, discretion requires high competency and professionalism on the part of an official and the ability to "seek resolutions to problems through recourse to interpretation of laws" (OECD 2017, p. 56). Experts point out that "the capacity of officials to seek resolve is stymied by concerns with the application of legislation" (OECD 2017, p. 56). If there is the possibility of transferring a citizen's appeal to another administrative body, so much the better; it becomes an optimal "decision" (Podoprigora 2013). For example, while the majority of appeals, on average around one-third, are registered on the central level, subsequently they are redirected to the local level. In 2016, although central bodies received 23% of all complaints, every third one was redirected to local government since, as officials argue, citizens frequently are not aware about the division of responsibility among public administration bodies. However, it may also be the case that citizens do not trust the ability of local officials to make an adequate and fair decision.

Despite clear advantages of the administrative appeal in terms of cost and time, its role is marginal and largely ineffective. A survey carried out in the 1990s showed that 73% of judges participating in the study thought that it is more effective to consider complaints in courts without preliminary administrative review (Gabbasov 2015). This is because the administrative complaint process, as argued, serves the purpose of avoiding judicial review of administrative actions. It is likely that a survey of citizens' positions in this respect could have shown an even higher incidence of negative attitude towards the internal complaint process (Gabbasov 2015; Policy 2019). A much later survey (n/d) conducted by the Ministry of Justice demonstrates the same attitude: the survey respondents trust courts more than the administrative review process. Administrative review, as observed, suffers from the lack of objectivity, bias, and ineffectiveness (Kuibeda 2010; Gabbasov 2014, 2015; IZ 2018; Policy 2019). Even so, it should be emphasised that courts also are inclined to take the side of public administration bodies as opposed to defending the weaker position of a citizen (Podoprigora 2012b, 2013). For example, in 2017, courts considered only 18,307 cases related to disputes with public administration bodies and made 10,370 decisions. Out of these, only 3510 (19%) were upheld while 37% of the cases were rejected. Apparently, according to a representative of Kazakhstan's Supreme Court, a similar trend was observed in 2018: 6724 cases were rejected, while 3826 were satisfied (Zakon.kz 2019).

Principles Underpinning Administrative Procedures

As for the use of principles of administrative procedures the situation is as follows. While each of the 11 principles is important in its own right, some of the principles in the list are questionable. For example, such principles as "[s]upport of state authority and prevention of actions that could discredit the Republic of Kazakhstan and be contrary to the interests of civil service", or "[v]ertical hierarchy of accountability of public administration bodies", or "[d]ivision of competencies and coordinated functions of state bodies and public officials", or "[e]fficiency and effectiveness" are internal civil service principles. The principle of "[s]upport of the state authority" paradoxically could, and does in practice, effectively override the principle of balance of interests of individuals and those of society and the state. The Soviet tradition of seeing the role of the state as guardian or protector of its citizens (and not as the state accountable to its citizens) continues in post-Soviet Kazakhstan. However, those principles of administrative justice that are recognised and practised in most of the advanced democratic administrative systems, as experts argue, are referred to vaguely or are not noted at all. These are the principle of prohibition of formalism; limitation of discretion; proportionality; and procedural fairness (Gabbasov 2012).

While many provisions of the LAP are duplicated in other laws including the LCC, there is divergence in wording and emphasis, and even differences in the list of general principles as opposed to those that are contained in other laws, rules, and regulations. For example, the LCC lists only six principles out of which the rule of law and prevention of red tape principles coincide with those in the LAP. In another example, the LAP specifies a duty to hear, although the Law on Special Social Services, which regulates provisions of services for a special category of citizens in difficult life situations, does not contain any provision that would adhere to this aspect of procedural fairness (Gabbasov 2014). In other words, there is no duty to hear and to be heard before deciding an issue in the Law on Special Social Services. If the law is silent on this procedural element, the professional civil servant will apply the LAP. However, this brings us to the vicious circle of the symbolic and superficial nature of this law.

If administrative procedures in Kazakhstan are considered to be secondary and unimportant, the same is true of principles of administrative law. In part this is because administrative law traditionally is considered one of the most conservative branches of public law in the Central Asian Soviet legislative tradition (Porokhov 2010; Deppe and Pudelka 2014). If the general principles are not set in the text as a norm, the likelihood that they will be applied in practice is very small. Furthermore, if these principles are applied, this could be viewed as frivolous administrative behaviour (Porokhov 2010). However, given that certain principles are actually outlined in the legal provisions, there could be some other reason why they are disregarded in practice.

The dominant principle upon which public officials rely is supremacy of law, which implies that administrative officials should act only within the framework of legal rules. Meanwhile, other important principles—including proportionality, fairness or equality before the law, and transparency—are used superficially or not used at all. As a result, not all circumstances of the case are investigated, and decisions are made on the basis of incomplete evidence (Kuibeda 2016; Sherstoboyev 2017).

Significant Changes, with More Needed

Experts in the field of administrative law for years were expecting the adoption of a comprehensive, consolidated Administrative Procedural Code that would displace ineffective law and would constitute an important step towards development of a law-governed state in Kazakhstan (Kuibeda 2016; Deppe and Pudelka 2014). After more than a decade of work, during which multiple drafts were produced, the administrative procedural code was finally adopted in June 2020 and will come into force on July 1, 2021.

Not withstanding the significance of automation of government services and a focus on the final result, the current developments in public administration will not be able to positively affect the quality of administrative justice in Kazakhstan without radical changes in the public officials' mindset: their work needs to ensure that the citizens' interests are served and they receive the benefits within parameters set by the legal framework. As experts emphasize, the aim is not just adoption of a decorative legal instrument that would serve to justify administrative decisions (Podoprigora 2015; 2010b). It is crucial to institute effective mechanisms for protection of citizens interests in disputes with state bodies and to provide guarantees that would allow individuals to participate in the process of administrative decisions that affect them. Furthermore, coupled with an exceptionally low level of legal administrative culture among the majority of participants of administrative procedures including citizens and officials (Gabbasov 2014), improvements would require political will to deal with procedural fairness in the relationship between citizens and public administration. Consider the following in the context of balance between results and the process.

In 2013, a national test resulted in 30% of higher ranking civil servants failing to show the most basic knowledge of administrative law (Emrich-Bakenova 2019). However, the resulting criticism focused not on the failure rate but on the substance of the tests, and on the need to abolish such kinds of tests since lack of knowledge of laws does not preclude a public administrator from being a good manager. As one expert emphasised, failure to show knowledge of the law should not cast doubt on the professionalism and decency of the administrator: "If this person is able to manage an entire sector of the economy or a region, what is the point of testing him on the knowledge of the law, especially if it is not relevant for his area of expertise?" (Ashimbayev 2013). What matters in the work of public managers is not "the letter of the law, but its quintessential value and meaning" (Ashimbayev 2013). However, what should be done if public managers fail to recognise the value of legal principles? In any event, the discussion boils down to a problem with the law and the importance of measuring actions of officials by final results.

Conclusion

This chapter has provided an overview of public service developments with the focus on issues pertaining to citizens' rights and interests in post-Soviet Kazakhstan. As the chapter argues, more is known, both in Kazakhstan and beyond, about achieved results to promote the country's quest to join the top 30 global economies by 2050 within the framework of the NPM than about administrative procedural reforms.

The chapter also argues that, despite assuming the role of a modern service-oriented state and, within that, the niche of a competitive provider of public services to customers, Kazakhstan's public administration nevertheless substantially demonstrates the same perspective on the relation between citizens and administration as in its Soviet past. While Kazakhstan was quick to adopt an international managerial approach to public administration, the stage of development of traditional law-based and procedure-oriented administrative culture was essentially skipped. Furthermore, the international best practice and models of administrative procedural fairness and principles are being taken on board very slowly. As public administration in Kazakhstan has become more complex, it has come to require stronger procedures and forms to facilitate disputes between citizens and the state.

In the light of all this, it would be important that the government takes action to:

- facilitate a balanced approach to public administration as a provider of public services with the role of a guarantor of citizens' rights;
- study the impact of technology on, and automation of, public services in the adoption of administrative decisions especially in the context of administrative fairness; and
- encourage public officials to consider administrative principles in their decision-making.

References

Ametistova, O., & Gabbasov, A. (2010). Administrative procedures. In R. Podoprigora (Ed.), *Administrative law: Textbook* (pp. 214–262) [in Russian]. Almaty: Tax Expert.

Ashimbayev, D. (2013, June 7). *About games and battles* [in Russian]. Panorama. Available via http://kazbio.info/?S=200. Accessed 16 February 2020.

Benish, A. (2014). Outsourcing, discretion, and administrative justice: Exploring the acceptability of privatized decision making. *Law & Policy, 36*(2), 113–133.

Bhuiyan, S., & Amagoh, F. (2011). Public sector reform in Kazakhstan: Issues and perspectives. *International Journal of Public Sector Management, 24*(3), 227–249.

Bizhikeyeva, M. (2019). *Feedback is required* [in Russian]. Available via https://kapital.kz/tehnology/79570/svyaz-dolzhna-byt-obratnoy.html. Accessed 23 January 2020.

Bovens, M., & Zouridis, S. (2002). From street-level to system-level bureaucracies: How information and communication technology transforming administrative discretion and constitutional control. *Public Administration Review, 62*(2), 174–184.

Brewer, B. (2007). Citizen or customer? Complaints handling in the public sector. *International Review of Administrative Sciences, 73*(4), 549–556.

CSA-AC. (2018). *Agency for Civil Service Affairs and Anti-Corruption. National report on state of civil service in the Republic of Kazakhstan* [in Russian]. Available via www.kyzmet.gov.kz. Accessed 27 January 2020.

Deppe, J., & Pudelka, J. (2014). *General administrative law in Central Asia—A brief survey*. Available via http://ruleoflaw.eu. Accessed 14 January 2020.

Emrich-Bakenova, S. (2019). Civil service development in Kazakhstan. In A. Farazmand (Ed.), *Global encyclopedia of public administration, public policy, and governance*. Switzerland AG: Springer Nature.

Gabbasov, A. (2012). *Administrative discretion and administrative justice in the Republic of Kazakhstan* [in Russian]. Available via https://online.zakon.kz/Document/?doc_id=31153141. Accessed 15 January 2020.

Gabbasov, A. (2014). *Theoretical and legal analysis of the law on administrative procedures of the Republic of Kazakhstan* [in Russian]. Available via https://www.zakon.kz/4595094-teoretiko-pravovoy-analiz-zakona-rk-ob.html. Accessed 17 December 2019.

Gabbasov, A. (2015). *Importance of preliminary procedures: Arguments for and against* [in Russian]. Available via https://www.zakon.kz/4729414-objazatelnoe-predvaritelnoe.html. Accessed 2 February 2020.

Gleeson, M. (2006). Outcome, process and the rule of law. *Australian Journal of Public Administration, 65*(3), 5–14.

Institute for Management Development (IMD). (2019). *IMD world competitiveness ranking 2019 versus 2018*. Available via https://www.imd.org/contentassets/b89907bec15a4d0a87316c212033f7cb/one-year-change-vertical.pdf. Accessed 24 November 2019.

IZ. (2018) *Review of conflicts before court review: Problems and perspectives*. Analytical report [in Russian]. Available via http://www.iz.adilet.gov.kz/sites/default/files/upload-files/otchet_28.11.2018.pdf. Accessed 19 November 2019.

Janenova, S., & Kim, P. S. (2016). Innovating public service delivery in transitional countries: The case of one stop shops in Kazakhstan. *International Journal of Public Administration, 39*(4), 323–333.

Janenova, S., & Knox, C. (2017). Civil service reform in Kazakhstan: Trajectory to the 30 most developed countries? *International Review of Administrative Sciences, 85*(3), 419–439.

Janenova, S., & Yesdauletov, I. (2017). Innovative public service reforms: What Kazakhstan can learn from Canada? *International Journal of Civil Service Reform and Practice, 2*(1), 32–48.

Kapital.kz. (2017). *Every civil servant should have the right to a mistake* [in Russian]. Available via https://kapital.kz/gosudarstvo/64945/bagdat-musin-u-kazhdogo-chinovnika-dolzhno-byt-pravo-na-oshibku.html. Accessed 21 January 2020.

Kassen, M. (2017). Open data in Kazakhstan: Incentives, implementation and challenges. *Information Technology & People, 30*(2), 301–323.

Kassen, M. (2018). Promoting public cooperation in government: Key drivers, regulation, and barriers of the e-collaboration movement in Kazakhstan. *International Review of Administrative Sciences, 85*(4), 743–762.

Kaufmann, W., & Feeney, M. K. (2014). Beyond the rules: The effect of outcome favorability on red tape perceptions. *Public Administration, 92*(1), 178–191.

Kazakhstan. (2018). Report on open government. *Evaluation of effectiveness of public administration bodies* [in Russian]. Available via https://evaluation.egov.kz. Accessed 17 December 2019.

Knox, C. (2008). Kazakhstan: Modernising government in the context of political inertia. *International Review of Administrative Sciences, 74*(3), 477–496.

Kuibeda, R. (2010). *Problems of inclusion of rules concerning administrative violations in the draft of administrative procedure code of the Republic of Kazakhstan* [in Russian]. The Legal Policy Research Centre (LPRC). Available via https://lprc.kz/. Accessed 19 December 2019.

Kuibeda, R. (2016). *Expert review of the draft of administrative procedural code of the Republic of Kazakhstan* [in Russian]. The Legal Policy Research Centre (LPRC). Available via https://lprc.kz/. Accessed 19 December 2019.

MacDonnell, V. (2015). The civil servant's role in the implementation of constitutional rights. *International Journal of Constitutional Law, 13*(2), 283–406.

Nazarbayev, N. (2013). *Meeting of the head of the state Nursultan Nazarbayev with civil servants of corps A in Astana* [in Russian]. Available via http://www.akorda.kz/ru/events/astana_kazakhstan/astana_other_events/segodnya-v-astane-glava-gosudarstva-nursultan-nazarbaev-provel-soveshchanie-s-gosudarstvennymi-sluzhashchimi-upravlencheskogo-korpusa-a. Accessed 29 October 2019.

Olsen, J. P. (2005). Maybe it is time to rediscover bureaucracy. *Journal of Public Administration Research and Theory, 16*, 1–24.

Organisation for Economic Cooperation and Development (OECD). (2017). *Decentralisation and multi-level governance in Kazakhstan*. OECD Public Governance Reviews. Paris: OECD Publishing. Available via http://dx.doi.org/10.1787/9789264280700-en. Accessed 12 September 2019.

Parliament.kz. (2019). *Nigmatulin: Quality of public services affects citizens attitude toward the state* [in Russian]. Available via http://www.parlam.kz/ru/mazhilis/news-details/id46837/1/1. Accessed 19 November 2019.

Podoprigora, R. (n.d.). *Issues in administrative law of the Republic of Kazakhstan* [in Russian]. Available via https://online.zakon.kz/Document/?doc_id=34634662. Accessed 23 January 2020.

Podoprigora, R. (2010a). Administrative law as a branch of public law. In R. Podoprigora (Ed.), *Administrative law: Textbook* (pp. 42–59) [in Russian]. Almaty: Tax Expert.

Podoprigora, R. (2010b). Introduction to administrative law. In R. Podoprigora (Ed.), *Administrative law: Textbook* (pp. 23–41) [in Russian]. Almaty: Tax Expert.

Podoprigora, R. (2012a). *Strategic issues in the development of administrative justice in Kazakhstan* [in Russian]. Available via https://online.zakon.kz/Document/?doc_id=34359522#pos=3;-253. Accessed 3 September 2019.

Podoprigora, R. (2012b). Administrative justice in Kazakhstan: Continuation of discussion. Available via https://online.zakon.kz/Document/?doc_id=31449474. Accessed 16 December 2017.

Podoprigora, R. (2013). Administrative procedures in Kazakhstan: Main issues [in Russian]. Available via http://www.zqai.kz/ru/node/12756. Accessed 16 December 2017.

Podoprigora, R. (2014). Administrative justice in Kazakhstan: History, development and perspectives [in Russian]. *Law and State, 4*(65), 46–56. Available via http://km.kazguu.kz/article/186. Accessed 14 December 2019.

Podoprigora, R. (2015). *Changes in administrative law of Kazakhstan will allow effective resolution of conflicts between citizen and state* [in Russian]. Available via https://cabar.asia/ru/roman-podoprigora-izmeneniya-v-administrativnom-zakonodatelstve-kazakhstana-pozvolyat-grazhdaninu-effektivnee-reshat-konflikty-s-gosudarstvom/. Accessed 16 December 2017.

Policy. (2019). *Draft of administrative procedural code of the Republic of Kazakhstan* [in Russian]. Available via http://sud.gov.kz/rus/node/50967. Accessed 18 January 2020.

Porokhov, E. (2010). Sources of administrative law. In R. Podoprigora (Ed.), *Administrative law: Textbook* (pp. 81–99) [in Russian]. Almaty: Tax Expert.

Prokuror.gov.kz. (n.d.). *Committee on the Legal Statistics and Special Accounts of the State Office of Public Prosecutor of the Republic of Kazakhstan* [in Russian]. Available via http://pravstat.prokuror.gov.kz/rus/o-kpsisu/deyatelnost-komiteta/analiticheskaya-informaciya. Accessed 21 November 2019.

Prokuror Kostanai. (2015). *Analysis of statistical data 1-OL. Consideration of citizens' and legal entities' appeals*. Kostanai region [in Russian]. Available via http://pravstat.prokuror.gov. Accessed 16 December 2019.

Shatov, V. (2012). *Problems of judicial review of normative legal acts of public organizations and officials* (example of judicial review of activities of tax agencies) [in Russian]. Available via https://online.zakon.kz/Document/?doc_id=37995768#pos=20;-31. Accessed16 December 2017.

Sherstoboyev, O. (2017). Administrative discretion and presumption of rationality [in Russian]. *Vestnik VGU, 4*, 74–83.

Shormanbayeva, A. (2012). *Freedom of movement in administrative law of the Republic of Kazakhstan*. Available via https://online.zakon.kz/Document/?doc_id=37995768#pos=20;-31. Accessed 16 December 2017.

Suleimenova, G., Kapoguzov, E., Kabizhan, N., & Kadyrova, M. (2019). Performance evaluation of the government agencies of Kazakhstan. *The NISPAcee Journal of Public Administration and Policy, XI*(2), 171–198.

Tokayev, K.-J. (2020). *Message of the President Kassym-Jomart Tokayev to the Government of the Republic of Kazakhstan* [in Russian]. Available via http://www.akorda.kz/. Accessed 14 February 2020.

World Bank. (2017). *Government effectiveness—Country rankings*. Available via https://www.theglobaleconomy.com/rankings/wb_government_effectiveness/. Accessed 11 January 2020.

Zakon.kz. (2019). Online news portal.

Zalnieriute, M., Moses, L., & Williams, G. (2019). The rule of law and automation of government decision-making. *The Modern Law Review, 82*(3), 425–455.

CHAPTER 6

Public–Private Partnerships in Kazakhstan: Evolution of the Government Policy and Reality of PPP Deployment

Shaimerden Chikanayev

Introduction

A number of policy documents indicate that over many years Kazakhstan has been showing political will to develop public–private partnerships[1] (PPPs). Not only did the government set a well-established legal (i.e. hard law) and institutional framework for PPPs, but also officially confirmed its commitment to develop the PPPs in its soft law, including in the *Kazakhstan-2050 Strategy* (Nazarbayev 2012) and the *2025 Strategic Development Plan*. By 2019, however, it became increasingly evident to all major stakeholders that PPP development in Kazakhstan was going in the wrong direction, as the PPP policy and governance proved to be

[1] This chapter does not discuss the definitions, features, and models of public–private partnerships. For the general understanding of PPPs, please see Mouraviev, N., & Kakabadse, N. (2016). Conceptualising public-private partnerships: A critical appraisal of approaches to meanings and forms. *Society and Business Review, 11*(2), 155–173.

S. Chikanayev (✉)
GRATA International Law Firm, Almaty, Kazakhstan

A. Koulouri and N. Mouraviev (eds.),
Kazakhstan's Developmental Journey,
https://doi.org/10.1007/978-981-15-6899-2_6

inefficient and have often been used to conceal public borrowing. The problems of the PPP mechanism not working properly became so acute that in 2019 the President of Kazakhstan, Mr Kassym-Jomart Tokayev, acknowledged that the whole concept of PPP in Kazakhstan had been discredited (Tokayev 2019).

However, in 2020, Kazakhstan is facing major economic and social challenges due to the impact of the COVID-19 pandemic and drop in the price of oil to a 20-year low, which has resulted in decreased economic activity, devaluation of local currency (tenge), reduced foreign direct investment, and shrinking state budget. This makes the need to attract investment in deteriorating public infrastructure even more acute, and the government is now under even greater pressure to find answers to hard questions about maintaining the volume and quality of public services. Therefore, Kazakhstan's government needs to take action in the near future to recalibrate the PPP policy and the way it is implemented.

This chapter examines PPP development in Kazakhstan from 1991 to the time of writing and reviews evolution of the PPP legislation and approaches used to develop a PPP-enabling environment in order to draw lessons for the country's future policy. This chapter describes evolution of the PPP concept in the country and provides analysis of gaps and deficiencies in the current PPP policy and legislation.

Since 1991 when Kazakhstan declared its independence, it has had to rely on the deteriorating infrastructure it inherited from the Soviet era. As a result, nearly every element of public infrastructure, such as roads, airports, water networks, and hospitals, required upgrading or expansion. Faced with gloomy growth projections and shrinking budgets after the collapse of the Soviet Union, Kazakhstan's government quickly recognised that it could use an effective method for attracting private investors and financing its growing infrastructure needs: PPPs. As the concept was novel for the country, the PPP definition and the scope of PPP activity has evolved significantly in the nation's legislation and legal literature since 1991. It is, therefore, useful to distinguish the following five stages of the PPP development in Kazakhstan:

- Stage One—PPP legal framework for foreign investors only (1991–1993).
- Stage Two—Lack of PPP-specific legal framework and "pilot" projects (1994–2005).

- Stage Three—The formation of legal and institutional frameworks (2006–2015).
- Stage Four—Active PPP deployment: the focus being on quantity (2016–2018).
- Stage Five—Adjustments to the PPP framework: stricter requirements (2019–present).

Each stage will be discussed in detail, followed by the conclusion.

Stage One—PPP Legal Framework for Foreign Investors Only (1991–1993)

What Was Done

Just seven days after it declared its independence, Kazakhstan adopted the law on concessions, which was the first PPP-related law in the country. The new legal framework has been proofed and applied in at least four infrastructure projects financed by the Development Bank of Kazakhstan and three concession projects supported by the national government.

Policy Goals at This Stage

The principal peculiarity of the first law on concessions was that at this initial stage of the PPP development a concession was viewed as an exclusive right or authorisation (i.e. de facto a licence) issued by the government to a concessionaire to engage in certain business activities. In other words, a concession agreement was some kind of an "administrative contract" (Popondopulo and Shevelyova 2015) that should not even be treated as a civil law agreement between equal parties because de facto it was a unilateral government authorisation (i.e. a licence) based on the provisions of public law (Moroz 2006).

Although the law on concessions was not industry-specific, it has been designed mainly for implementation of projects related to utilisation of natural resources. As at that time oil and gas were viewed as the only possible drivers of the national economy, it was well justified that the law was tailor-made for the oil and gas projects.

Importantly, the law on concessions could be used only by foreign investors as the law stipulated that a concession could be granted only to foreign legal entities and/or citizens. When the Soviet Union disintegrated, there was no market economy in Kazakhstan and no local businesspeople who could act as concessionaires. It is, therefore, natural that at the dawn of Kazakhstan's independence, attracting foreign direct investment was the main PPP policy goal aiming at bringing in money and technology for exploration and extraction of the country's oil and gas reserves.

Assessment

The scope of the PPP concept during Stage One was limited to the concession form. It was understood narrowly as an exclusive right to be awarded to an investor (in some countries, a concession is understood in the same way as exclusive right, although in others it is viewed as one of many types of a PPP agreement; see, for example, Tuktarov and Dubinchina 2016). Stage One represents a time when the government was not yet ready for any partnership with business on equal terms and, therefore, the whole PPP model (i.e. a licensed-type concession agreement based on public law provisions) was designed to enable the state to dictate its terms of partnership to the business. It is no surprise, therefore, that just a handful of projects had been implemented under this novel legal framework. The first law on concessions was already deemed invalid by April 1993.

Implications for Further Policy Development

Rapid transition to a market economy in the 1990s exposed deficiencies of the licensed-type concession agreement under the law on concessions, which proved to be a hindrance for attracting private investment. The government learned its lesson and all new types of PPP-related agreements it introduced thereafter have been based on private law and not public law provisions. Instead of the power–subordinate relationship that existed under the first law, all PPP-related types of agreements introduced in Kazakhstan after 1993 have been based on the "equality of parties" principle.

Another lesson Kazakhstan learned from this stage is that PPP-related laws should be tailor-made for developing public infrastructure (*economic* infrastructure and *social* infrastructure—see Yescombe and Farquharson 2018) and not for the subsoil utilisation projects. This materialised in

2006 when a new Concession Law specifically excluded subsoil use operations from its application (Law of the Republic of Kazakhstan 2006). From 2006, the legal term "a concession agreement" could not be applied to any type of the subsoil use contracts (Bassin et al. 2004), whereas the Law on Public-private partnership (the PPP Law) (Law of the Republic of Kazakhstan 2015) adopted in 2015 includes the provision that the term may be applicable to projects involving subsoil use but only in conjunction with the provisions of the Subsoil Use Code.

Finally, unlike the first law on concessions, all PPP-related laws thereafter have been applicable to both domestic and foreign investors, as the government realised the importance of treating all investors equally.

Stage Two—Lack of PPP-Specific Legal Framework and "Pilot" Projects (1994–2005)

What Was Done

From 1993, when the first law on concessions stopped working, to 2006, when a new law on concessions came into effect (i.e. for about 13 years), there was no PPP-specific legislation in Kazakhstan. Despite the absence of any specific laws on PPPs or concessions, a number of agreements have been executed during this time between public law entities (i.e. the Republic of Kazakhstan, its regions or cities) and investors, based on PPP principles and relying solely on the general principle of freedom of contract in Kazakhstan's civil law.

Privatisation of the country's electrical power and heating infrastructure using the PPP mechanism, in the second half of the 1990s in particular, triggered interest amongst foreign investors. In 1997, the Belgian company Tractebel obtained the "concession" for the electricity, heating, and gas supply in the city of Almaty (the nation's then capital), as well as the licence for the construction and operation of gas pipelines. The American company AES also received concessions for the provision of electricity and heating in the same year and invested in some of the country's largest power plants, including the 4000 MW Ekibastuz-1 coal-fired power station. However, the foreign investors quickly faced high-profile disputes with the government. Disagreements regarding tariffs and a corruption scandal resulted in the early termination by Tractebel of its investment in the beginning of the 2000s. Similarly, amid regulatory disputes with the government, AES sued Kazakhstan before investment arbitration and divested its largest energy assets in the country in order to gradually close down its Kazakhstan-based business (Boute 2019).

Nonetheless, the other two infrastructure projects were launched as concession agreements in 2005, despite the absence of a specific legal framework. These projects are usually viewed by the government as pilot PPP projects (although they were not the first ones as there were others, as discussed above):

1. Construction and operation of the inter-regional power transmission line "North Kazakhstan – Aktobe Region" in the Aktobe region (a concession for 25 years with a concessionaire Batys Transit).
2. Construction and operation of the railway line "Shar Station – Ust-Kamenogorsk" in East Kazakhstan region (a concession for 23 years with a concessionaire Doszhan Temir Zholy) (for details of this project, see Eurasian Economic Commission 2018).

Both concessions performed below expectations and could hardly be viewed as successful PPP projects. In part, this was caused by poor structuring of these projects. In addition, there was apparent lack of proper balance of risk sharing between an investor and the state. For example, the railway has seen less freight than expected, which had a serious negative impact on the project's revenue. In the power transmission project, although several enterprises were expected to use the power line, they were not working or operating below full capacity. Both concession projects managed to raise debt financing on the domestic stock exchange by issuing infrastructure bonds secured by the state sureties. These state sureties issued their guarantees under Kazakhstan's law. However, their guarantees proved to be unreliable means of security in practice. Doszhan Temir Zholy (the operator of the "Shar Station – Ust-Kamenogorsk" railway line project) declared default on its infrastructure bonds in 2008, and many of Kazakhstan's pension funds that acquired these bonds failed to enforce their claims against the Republic of Kazakhstan as a surety. After some time, however, in the light of looming loan defaults and risk of bankruptcies of the companies involved, the government stepped in and rescued both pilot PPP projects by purchasing part of the shares from private investors.

Policy Goals at This Stage

Foreign participation in the privatisation of electrical power and heating generation and supply was an important component of the government's successful policy to address a severe debt crisis that hit Kazakhstan's energy sector at the end of the 1990s (Boute 2019). The government expected that attracting foreign investors by way of PPPs and privatisation would (i) facilitate the restructuring of the debts accumulated by the state-owned operators of utilities, (ii) bring in much-needed capital and technology for the modernisation of the energy sector and (iii) reduce the budget expense on the power and heating generation.

In addition to upgrading public infrastructure, the government also hoped to use PPPs for developing the domestic capital market by introducing a new financial instrument—infrastructure bonds—that were expected to become a lucrative and, importantly, secure asset for investment by the nation's pension funds, insurance companies, and other large firms.

Assessment

Stage Two captures government–business relations in Kazakhstan at a time when the government tried to use PPPs and privatisation in the most flexible and unregulated manner to do much-needed reforms in the energy and utilities sectors and swiftly attract foreign and domestic investment. Yet, the government learned the hard way that there are not really any shortcuts to forge a lasting partnership with the private sector.

The second stage, therefore, presents particular interest for legal research because any form of cooperation between the government and business was, de facto, recognised as PPP and all projects implemented as such during this time have been structured and launched in the absence of a PPP-specific legal framework. The absence of specific legislation means that the legality and enforceability of the so-called concession agreements remain questionable. De jure these agreements could not be qualified as concession agreements as provided by the 2006 Concession Law but should be qualified as so-called innominate contracts (*contractus innominati*). In particular, one of the grey areas is the legal capacity of public law entities (i.e. the Republic of Kazakhstan, its regions, and certain cities) to enter commercial contracts, if such commercial contracts are not specifically provided by Kazakhstan's law (i.e. if they are not so-called nominate contracts from Kazakhstan's civil law perspective) and if Kazakhstan's law does not specifically authorise relevant state bodies to sign such contracts on behalf of the respective public law entity. In the absence of such specific regulation, transparency of the process related to selecting investors, as well as legal the authority of relevant state bodies (i.e. the national government or regional and local executive bodies) who signed concession agreements on behalf of relevant public law entities, also remains an open issue.

The lack of a transparent private partner selection procedure and the absence of a solid legal framework for structuring PPPs during Stage Two of PPP development in Kazakhstan, unsurprisingly, left too much room for the discretion of government officials and opportunities for corrupt practice. During this stage, many PPP tenders were rigged and lacked competition, being largely controlled by favoured insiders with political connections to the government bodies. It is of no surprise, therefore, that most litigations and corruption scandals surrounding PPP projects occurred during this time (e.g. a dispute between CCL Oil Ltd and Kazakhstan's government in relation to concession agreement executed in 1997 with regards to the Pavlodar oil refinery is just one more example, in addition to AES and Tractebel cases discussed in this section).

Implications for Further Policy Development

The principal lesson from Stage Two is that in order to attract investors on a long-term basis, specific PPP legislation should be adopted to make sure that the private partner selection process is transparent and efficient, whereas authorities of the relevant state bodies should be clearly stipulated in legislation.

The government also realised that the exceptionally low tariffs that were used in the utilities and energy sectors did not enable the utilities companies to recover their costs and attract foreign investors and did not allow viable PPP projects to be launched without additional payments from the state budget. Raising the tariffs for utilities is typically considered a politically unpopular action; it is a sensitive issue that could trigger social unrest. Therefore, the government has been, and still is, reluctant to attract foreign investors and initiate any PPP projects in the utilities sector. This calls for a fundamental reform of the tariff-setting methodology used by the natural monopolies in general and in utilities in particular.

Private investors always seek clarity about the government's commitment to adopt a tariff-setting policy that would ensure the financial viability of the contract, although a project often has to be accompanied by the transparent subsidies if the government decides that not all consumers can afford to pay cost-recovery tariffs.

Stage Two also showed that Kazakhstan needs more effective PPP policy instruments for the implementation of projects in the utilities sector. For example, it might be a PPP model, specifically provided by law, that allows payments to a private company by both the government and final users or allows shadow tolls (i.e. when the government pays the tolls in place of the final user) to guarantee certain revenue to a private party for a pre-specified volume of public services (Mouraviev and Kakabadse 2017).

Finally, because many pension funds burned their fingers on infrastructure bonds during this stage of PPP development, this type of bonds—used as a mechanism for raising debt financing for PPPs—has not been used in Kazakhstan ever since. It shows that the government largely lost interest in the domestic capital market as a source of financing for PPPs and instead opted to attract the necessary funding mainly from international institutions.

Stage Three—the Formation of Legal and Institutional Frameworks (2006–2015)

What Was Done

Although Kazakhstan received some experience of PPP-based projects during stages one and two, successful deployment of new partnership projects would not be effective without comprehensive statutory legislative package on PPP procurement. During Stage Three, Kazakhstan managed to put in place well-structured legal and institutional frameworks for launching PPPs at both the national and regional levels (European Bank for Reconstruction and Development (EBRD) 2018; Rivadeneira and Garin 2019).

In 2006, the new Concession Law was adopted, which already in 2007 enabled execution of the now specifically recognised concessions (i.e. a concession agreement now became a nominate contract from Kazakhstan's civil law perspective). These were concession agreements on construction and operation of the passenger terminal of the International

Airport in the city of Aktau; the Yeraliyevo-Kuryk railway line (construction and operation); and electrification of the Makat-Kandyagash railway line. This was followed by execution in 2008 of the concession agreements on construction and operation of the gas turbine plant in the city of Kandyagash (in Aktobe region) and the railway line Korgas-Zhetygen (Mouraviev et al. 2012). However, the then largest PPP project that was expected to pave the way for more partnerships with foreign investors and lenders was the Big Almaty Ring Motor Road concession project (a toll road around Almaty), also known as BAKAD. The Ministry of Industry and Infrastructure Development with assistance of the International Finance Corporation as its advisor put the BAKAD project out to tender under the Concession Law framework at the beginning of 2015, although this long-delayed project was first conceived as early as 2008. The BAKAD 20-year, build-transfer-operate (BTO) concession agreement has been signed in February 2018 by a consortium (comprising South Korea's SK E&C, Korea Expressway Corp and Turkey's Alarko and Makyol) acting as a concessionaire. The project offers an enticing risk/reward combination. While construction cost amounts to USD 540 million, the total project cost is estimated at USD 750 million. While Kazakhstan's government agreed to assume all traffic risk, toll revenue will be collected by the concessionaire and transferred to the government straight away. In turn, the government will be making a front-loaded availability payment to cover investors' capital expenditure, plus will be paying interest during the first 10 years of operation to mimic the debt repayment schedule. At the same time, the investors' operating expenses, replacement capital expenditure, taxes, cost of capital and other costs will be covered throughout the whole operation period, to the end of the project. The government will also mitigate currency fluctuation risk and compensate for any depreciation of tenge (Kazakhstan's national currency) against the US dollar above 5%. Even though the BAKAD concession agreement was signed back in February 2018, it took another two years to reach financial close when all relevant stakeholders, including EBRD, managed finally to agree, in February 2020, on project financing (Newsworld 2020). There is high expectation for success of this project because if BAKAD will not be implemented for any reason, foreign investors, lenders and international financial institutions (IFIs), that so far supported Kazakhstan's government PPP initiatives, may lose trust in government's commitment to engage foreign partners in any PPPs.

Importantly, as of April 2020, out of all above-mentioned concessions only one has been working successfully (i.e. a facility has been constructed and is already functioning)—the passenger terminal of the Aktau airport. It means that the law on concessions is in effect already for 14 years, although only one concession project has been successful under its legal framework (Kazakhstan Public–Private Partnership Centre 2020).

Concessions lack popularity partly owing to excessive regulation by the relevant legislation and because the process for granting a concession in Kazakhstan is bureaucratic and time-consuming. To address this issue, the government throughout Stage Three tried to improve the Concession Law by introducing numerous changes. For example, upon its adoption in 2006 the law allowed concessions to be structured only as either BTO or design-build-transfer-operate (DBTO) projects (i.e. upon completion of the construction phase, ownership of the relevant concession facilities was required to be transferred to the government). This statutory limitation proved to be an unnecessary obstacle for concession projects, and investors often required greater flexibility. The Concession Law, therefore, later has been amended to allow all possible PPP models. Nowadays, a concession agreement may include an option for a private partner to keep ownership of the concession facility upon project completion. Currently, all available PPP schemes may be structured.

In addition to the Concession Law, Kazakhstan also adopted the Project Finance Law in 2012 when the project finance concept was introduced into national legislation. There is now a solid basis for the structuring of PPP schemes on the principles of project finance. So far, however, project finance in the strict sense of the term (i.e. where financing is based on the performance of the project itself) under the Project Finance Law has not yet been tested in Kazakhstan.

In addition to the legal framework, during Stage Three, the government managed to form quite a sophisticated institutional framework for PPPs (Rivadeneira and Garin 2019). For the national-level concession projects in their sector (e.g. the Ministry of Healthcare for hospital projects), the line ministries perform the role of a contracting authority (de facto on behalf of the Republic of Kazakhstan). The line ministries decide on the delivery mode for investment projects and develop documentation for PPPs. Local governing bodies (*akimats*) perform the role of a contracting authority (de facto on behalf of a respective region or a city) for all local-level concession projects. The state body that is procuring the PPP (i.e. a contracting authority) is the institution ultimately responsible for the project. The contracting authority is responsible for the preparation, negotiation, and administration of the contract, and for monitoring and evaluating contract performance during the construction and operation phases of the project. This authority, therefore, carries the ultimate responsibility for the PPP contract and its implementation (Rivadeneira and Garin 2019).

In 2008, Kazakhstan established its specialised unit called Kazakhstan Centre for Public–Private Partnership (the PPP Centre) located in the nation's then capital city, Astana, (renamed Nur-Sultan in 2019) to facilitate PPP projects. The centre is charged with the economic evaluation of proposed PPPs, including concessions and assistance with project preparation, such as evaluation of tender documentation, bids of potential private partners, and drafts of the concession/PPP agreements. Later, in 2014, the PPP Centre set up the Kazakhstan Project Preparation Fund LLP (KPPF) as a joint venture with the National Management Holding Baiterek which is an organisation managing a large number of the country's companies. The KPPF promotes infrastructure development in Kazakhstan through providing services for structuring and supporting infrastructure projects including those that make use of the PPP model. The KPPF can assist with drafting tender documentation, a concession or a PPP agreement. It also could be involved in negotiating with potential investors and concessionaries, as well as working with the government organisations regarding securing approval for the project documentation. The KPPF, in general, serves as an external independent adviser to the national government and sometimes to the local authorities. Finally, nowadays, almost every region (*oblast*) in Kazakhstan has its own PPP Centre responsible for the development of PPP in a respective region. Furthermore, a few line ministries in Kazakhstan have created their own units (departments) for developing PPP projects (Rivadeneira and Garin 2019).

Kazakhstan also managed to build up a diverse pool of different types of lenders that could finance PPP projects. Firstly, it is the Development Bank of Kazakhstan, a government-owned entity responsible for financing infrastructure projects including PPPs. Secondly, domestic commercial banks could be interested in financing PPPs and there are already small-scale projects financed by the commercial banks, despite lacking appetite and capacity to engage with large projects. Finally, most of the national-level PPP projects so far relied on IFIs. IFIs strongly support PPP development and their assistance to Kazakhstan's government is not limited to lending, but also includes rendering international expertise in managing PPPs. The Asian Development Bank (ADB), the EBRD and the International Finance Corporation (IFC) are especially active in Kazakhstan. Quite often they help the government by serving as advisors regarding PPP structuring or by bringing in international and local consultants in order to improve PPP-related legislation.

As for the PPP policy in Kazakhstan, it is mainly led by the Ministry of National Economy as an authorised state body that is assisted by the PPP Centre. The ministry sets the guidelines, drafts PPP legislation and has an overarching monitoring role with regards to PPPs. In addition, the Ministry of Finance, sectoral ministries, the local authorities and the National Chamber of Entrepreneurs (*Atameken*) also play an important role in the development of the PPP policy in the country.

Policy Goals at This Stage

In 2010, the government launched the Programme for Accelerated Industrial-Innovative Development of Kazakhstan in 2010–2014 (the Industrialisation Programme). It envisaged the implementation of at least 15 PPP projects per year totalling over KZT 900 billion (about USD 6 billion) by 2015. In furtherance of the Industrialisation Programme, in 2011 the government approved the Programme for Development of Public–Private Partnership in 2011–2015 (the PPP Programme). The expectation was that simultaneously the Ministry of Economic Development and Trade would facilitate the creation of an appropriate legislative and regulatory framework for implementation of infrastructure projects using PPP mechanism. Stage Three in the PPP development was, therefore, the first time when a national PPP policy document (the PPP Programme) was promulgated. Its major goal was to create specific legislative and institutional frameworks for PPPs. This goal has been successfully accomplished (as discussed above in the section What was done [Stage Three]). The government, however, fell short of achieving another goal set in the PPP Programme, namely to implement twice as many PPP projects by 2015 as compared to 2008. In reality, no projects have been implemented during this time, although a few concession projects, including BAKAD, have been announced during Stage Three.

The PPP Programme declared education, healthcare, toll roads, and housing and public utilities as priority sectors for PPPs. However, it seems that in fact the major goal during Stage Three was to attract foreign investment and implement about ten large-scale PPP projects, with a total value of more than USD 4 billion, in the transportation sector and, therefore, create the PPP projects that would serve as a benchmark (Matayev 2014).

To summarise, the PPP Programme as de facto national PPP policy proved to be an effective tool for PPP development as the government set clear objectives and targets and established the priority sectors for partnerships. This matched the expectation of foreign investors and other interested parties. Unfortunately, the practice of promulgation of similar national PPP policy documents did not continue after the PPP Programme had ended in 2015. It is not surprising, therefore, that the Asian Development Bank's PPP Monitor notes a lack of formal, published PPP guidelines and the list of projects in the pipeline as one of the impediments to PPP development in Kazakhstan (ADB 2017).

Assessment

During Stage Three of PPP development, the understanding of a PPP was narrowed to simply mean a concession (Mouraviev and Kakabadse 2017). In the 2006 Law on Concessions, a concession is defined as activity aimed at construction (or renovation) and operation of a facility, performed at the expense of a concessionaire or with co-financing by the grantor. The legislation, therefore, allows the public sector partner to provide additional forms of support to a concessionaire. The government can (and often does) provide financial support to a project, although the legislation limits this support to the total value of the completed project. The concession law includes the concepts of an *availability payment* to ensure continuous operation of a concession facility and *state subsidies* as both an additional source of income for a concessionaire and a tool to reimburse its expenses. The availability payment includes outlays from the government budget as compensation of concessionaire's certain investment and operational expenses and, if applicable, any service fees for trust management of the government-owned property (i.e. a concession facility) or lease payment paid by the government for the use of a concession facility owned by a concessionaire. Importantly, the Concession Law provides possibility to obtain an availability payment only for concession projects that have been categorised as "socially important", such as kindergartens. The Kazakhstan's definition of a concession, therefore, is different from traditional understanding of a concession in Western Europe and international legal literature (UNECE PPP Assessment Report 2013). In the traditional concept, a concession is limited to a special type of a PPP in which a private sector entity is allowed only to charge the general public for using the concession facility ("users pay" model) but is not able to claim any compensation from the state budget (Yescombe and Farquharson 2018).

Unlike the 1991 concession law that categorised a concession agreement as an administrative contract, the 2006 Concession Law classifies a concession agreement as a special type of a private law contract. It means that any concession agreement is a nominate contract that must have the following two elements as a subject matter: (i) creation/reconstruction of a concession facility and (ii) operation of this facility in accordance with its designated use. For instance, if the concession facility is a school, a concessionaire must deliver the core service by providing educational services (i.e. a "limited service" project under which the services provided by the private partner during the operations phase do not include educational services, but are limited to facility maintenance and management, cannot be structured under concession agreement). Another practical implication of categorising a concession agreement as a nominate contract is that it has its own specific regulation that is different from non-concession PPP agreements. For example, Kazakhstan's Tax Code provides specific/preferred tax regulation only for the concession agreements and not for any other types of PPP agreements.

Importantly, a concession facility can only be a property that may be recognised as the so-called social and vital infrastructure facility (i.e. a facility used for meeting the public needs, the securing of which is imposed on state authorities by Kazakhstan's law). The Concession Law, therefore, cannot be used for the construction of, for example, a fertiliser plant as it is unlikely to be considered a facility used for meeting public needs, the securing of which is imposed on state authorities. Unlike the Russian concession legislation, Kazakhstan's Concession Law does not provide an exhaustive list of the property types that could be objects of a concession agreement. However, in general, only immovable and movable property may be a concession facility in Kazakhstan (Tuktarov and Dubinchina 2016).

It is worth noting that there were two drafts of the concession law submitted for consideration of the Parliament in 2006. One draft was prepared by the Ministry of the National Economy and the other draft was prepared by civil law scholars led by renowned academic Professor MK Suleimenov. Although the law is a blend of the two drafts, unfortunately, the final version was mainly based on what the ministry proposed (Beisembinova 2011). This explains the gaps and ambiguous provisions that still exist in the concession law. For example, one of the major reasons why it took more than two years to reach a financial close with the lenders of the BAKAD project was confusion regarding who was the concession grantor. The 2006 concession law states that only authorised state bodies (e.g. the Ministry of Industry and Infrastructural Development, which signed the BAKAD concession agreement) can act as a concession grantor and enter into concession agreements, but this contradicts Kazakhstan's legislation because state bodies are not subjects of civil law and, therefore, they do not have any assets and cannot enter into any civil law agreements on their own behalf (i.e. strictly speaking state bodies can act and sign a concession agreement on behalf of the public law entities, e.g. the Republic of Kazakhstan or its region or a city, but not on their own behalf as legal entities). From the bankability perspective, it makes significant difference if the grantor is the Republic of Kazakhstan that has many assets, or just a ministry with effectively no assets at all to meet its contractual obligations.

Implications for Further Policy Development

At Stage Three, the principal learning point is that developing a policy document and setting legal and institutional frameworks are not enough for creating a PPP-enabling environment. Despite forming advanced legal and institutional PPP frameworks and promulgating a policy document (the PPP Programme), there was an apparent lack of success in PPP implementation during Stage Three. It seems that the country lacked the critical element that is required for successful PPP implementation: high-level political support that would overcome all kinds of resistance to private sector involvement in public service delivery. Many officials (in the line ministries and regional and local governments) at that stage did not see powerful incentives for their respective organisations to embark upon PPP implementation that back then was a novel, quite complicated and time-consuming mechanism compared to the traditional public procurement process. There is no surprise, therefore, that during Stage Three only one concession project (the passenger terminal of the international airport in Aktau) has been successful.

Stage Four—Active PPP Deployment: The Focus Being on Quantity (2016–2018)

What Was Done

The end of year 2015 became extremely fruitful in terms of making changes to the PPP legislation as the new piece of legislation—the PPP Law—and all necessary by-laws have been adopted (Law on Public–private Partnership 2015). Importantly, Kazakhstan's president in his address to the nation on 30 November 2015 specifically urged the use of PPP for economic development, which provided a powerful impetus to all officials and extended much-needed high-level political support for PPP implementation. Having established a solid legal and institutional basis for the PPP projects and a clear message from the president to implement PPPs in all sectors of the economy, at the end of 2015 the national government officially confirmed that from then on each region in Kazakhstan would have to meet a new target: to implement at least five PPP projects per year. The government has also set similar targets for the number of PPP agreements that line ministries must sign on an annual basis. It also instructed line ministries and regional authorities to develop a road map by March 2016 for PPP formation in their respective sectors and regions with the specific target indicators.

As a result, soon thereafter two contracts have been signed under the framework of the new PPP Law. These were the national-level PPP projects: a contract to build a checkpoint and a transport and logistics centre in the international transit corridor, "Western Europe – Western China", and a contract to design and operate an automated information system, "Kandelik" (Rivadeneira and Garin 2019). The checkpoint and a transport and logistics centre project were executed as a PPP agreement in 2017 without a public tender as a result of so-called direct negotiations with a domestic investor—LLC Eurotransit Terminal—and was financed by the Eurasian Development Bank. This was the first national-level PPP agreement and the only arrangement that had two public sector partners (two ministries) as counterparties of the agreement. The interpretation of the law, however, suggests that de facto there is only one public partner: the Republic of Kazakhstan represented by its two state bodies (ministries).

However, the regional level, rather than the national level, saw a rapid increase of the number of executed PPP agreements during Stage

Four. At this stage, half of the projects have a maximum investment not exceeding KZT 100 million (about EUR 250,000). Furthermore, 80% of the projects have a total value below KZT 500 million (EUR 1.25 million) (Rivadeneira and Garin 2019). Poor PPP project selection and preparation by the local authorities (mainly owing to a lack of competence in the regions and/or lack of relevant officials that were ready to take on the responsibility of making the necessary decisions) inevitably resulted in defaults, litigations, and the invalidation of many inadequately structured PPP agreements (some of which did not comply with the PPP legislation) and tenders during Stage Four (SZP 2019).

Still, the largest project at the local level is the Almaty Light Rail Transit (LRT) PPP, although there is confusion regarding who a public sector partner might be and against what assets the private sector partner might have legal recourse. Systematic interpretation of the 2015 PPP Law suggests it is the city of Almaty (rather than the Republic of Kazakhstan) and, therefore, a future private partner of the Almaty LRT project may have legal recourse only against the municipal budget of Almaty, but not the state budget of Kazakhstan. The city government is prepared to give a subsidy to cover 50% of the capital cost and retain ridership, but the city expects to receive all ticket revenue. The private sector partner would be paid an availability payment to cover its investment and operating costs. The Almaty LRT PPP project is also a good example of the public sector partner unexpectedly changing its mind in the middle of the tender process. Although four international consortia successfully passed the first stage of the open tender and in 2019 entered the second stage, the newly appointed mayor of Almaty currently proposes two options: (1) to abandon a PPP model for a light rail and instead build and operate the LRT system by the public sector organisations as, presumably, this option offers better value for money or (2) to abandon the project altogether. It is likely that a change of plans is triggered by doubts of whether it is feasible to attract private investors without very high costs or the central government guarantees for potential liabilities. Furthermore, implementing the project via public sector organisations, rather than the private firms, could save hundreds of millions of dollars.

Policy Goals at This Stage

During Stage Four, the government pursued two goals in furtherance of its new PPP policy developed by the Ministry of the National Economy in collaboration with the PPP Centre. The first goal was to establish a new legal framework under the PPP Law that should have provided for much broader understanding of the PPP concept. A new framework was expected to be more flexible compared to the one formed under the 2006 concession law (i.e. the new PPP Law did not intend to replace the Concession Law; rather, it aimed at providing a more flexible and robust solution for PPPs) (Rivadeneira and Garin 2019). The second goal was to implement many small-scale and simple (service-type) PPP projects under the PPP Law at the regional level, rather than just a few large-scale infrastructure projects at the national level under the Concession Law, as was the case of the PPP policy during Stage Three. To meet these goals, the government pushed for decentralised PPP preparation, approval, and implementation, so that the regional and local authorities would have more power and could implement the regional-level PPP projects faster, without a need to get any approvals from the PPP Centre or the national ministries. Supported by the PPP Law, the new policy has led to a sharp increase in PPP formation at the regional level, with 15 PPP agreements signed in 2016, 166 in 2017, and 304 in 2018 (Atameken 2019). These regional projects have been focused on the social infrastructure, including kindergartens and primary schools (54% of the total number of executed PPP agreements), primary healthcare centres (21%), and sports and recreation facilities, such as halls for martial arts (14%). Many of the projects developed under the PPP Law during this stage took the form of service contracts, such as street cleaning or waste processing, with limited capital investment (Rivadeneira and Garin 2019).

Assessment

The PPP Law significantly broadened the PPP concept by viewing a partnership as a form of cooperation between the public partner and a private partner that corresponds to the features defined by the law. These features include (i) relations between the public and private sector partners in the form of a PPP agreement, (ii) medium-term or long-term PPP project implementation (from 3 to 30 years), (iii) joint participation of the public and private sector partners in project implementation, and (iv) resource sharing by the partners to facilitate project implementation. A very broad definition of a PPP enables the launch of projects in any sector of the economy. Therefore, PPP facilities under the 2015 PPP Law, unlike the 2006 concession law, do not necessarily have to be used for satisfying the public needs, the securing of which is imposed on state authorities (e.g. the construction of a fertiliser plant or subsoil utilisation project are now eligible for implementation under the PPP Law).

The law enlists possible types of a PPP agreement, including a concession agreement, trust management of state-owned property agreement, and a life cycle contract; however, the list remains open, so it is possible to enter into other agreements that align with the PPP features. Logical and systematic interpretation of the PPP Law, however, suggests that in fact under the current law there are only two types of a PPP agreement: a concession agreement and a non-concession agreement.

- A concession agreement (for the purposes of the civil law) is a so-called *nominate contract* with its own specific legislation and even tax regulation. This means that a concession agreement should not be considered a type of a PPP agreement and, therefore, should not be subject to regulation by the PPP Law.
- A non-concession PPP agreement (for the purposes of the civil law) is a so-called *mixed contract* (i.e. a contract that contains the elements of different types of a contract provided by the law). A non-concession agreement also has its own special legislation and features.

This means that Kazakhstan's PPPs in effect have two forms: (1) a concession and (2) a non-concession PPP, which both have different legal frameworks.

The subject of a PPP agreement in the form of a mixed contract is not clearly defined by the PPP Law because it can vary depending on the types of contract a particular PPP agreement consists of. However, in general, it can be related to any type of activities including construction and/or developing infrastructure and/or rendering services. Furthermore, a PPP agreement is not only a mixed contract, but also a nominate contract, which means that a PPP agreement should be considered concluded only when relevant parties agree on all essential terms of the PPP agreement as stipulated in the PPP Law (Chikanayev 2017).

Most academics and businesspeople view a PPP as a contractual agreement (Mouraviev and Kakabadse 2017). However, a PPP acquired specific legal meaning under Kazakhstan's law during Stage Four and, therefore, the PPP concept in Kazakhstan nowadays covers only two specific contractual arrangements: (1) a concession agreement and (2) a PPP agreement. A wide variety of other contractual arrangements involving the public and private sector partners (e.g. trust management of state-owned property agreement executed under the State Property Law), executed not in accordance with either the 2015 PPP Law or the 2006 Concession Law, therefore, cannot be qualified as a PPP.

Legislation also failed to make clear how these two laws are linked and what distinctive features of the concession law make it, in certain cases, a preferred option when compared to the PPP Law. Nevertheless, the PPP Law currently has a number of provisions that are either unclear (and the approach to the application thereof must first be tested in courts) or too onerous to the business (Chikanayev 2015). This is why for the time being the concession model remains the most popular option for IFIs and foreign investors in Kazakhstan.

Implications for Further Policy Development

The evidence suggests that many local (city) and regional authorities have been improperly using the PPP concept during Stage Four in order to attract private investors or as a cover for irregular procurement practices or public borrowing. For example, under the PPP Law, a private sector partner may be selected through so-called direct negotiations (i.e. without an open tender) in certain cases as stipulated in the law. The problem is that, in practice, the direct negotiations method might be misused by the regional and local governments and affiliated business entities as a loophole that allows awarding a contract without competition. The fact that more than 50% of the PPP agreements executed by the end of 2018 were procured through direct negotiations (i.e. without an open tender) raises serious concerns. During Stage Four, there was also a rapid increase in the execution of different agreements that were called PPP agreements but, strictly speaking, should not be treated as such, as they are just imitations of a PPP (i.e. these agreements often do not correspond to the features and principles of the PPP as prescribed by the PPP Law and discussed in the Assessment section above) and should be requalified as public procurement contracts. This has caused confusion and uncertainty regarding PPPs, which was driven mainly by the swiftly formed negative perception of how PPPs develop in Kazakhstan. By the end of Stage Four, therefore, all relevant stakeholders expected the government to roll out changes in the PPP policy and legal and institutional frameworks to curb the growing government debt risk and prevent the misuse of fiscal funds by the regional and local authorities.

Stage 5—Adjustments to the PPP Framework: Stricter Requirements (2019–Present)

What Was Done

As the national-level PPP and concession projects in Kazakhstan were put under the microscope of many controlling state entities, including the PPP Centre, in an attempt to have a truly competitive and transparent PPP selection process, the regional- and local-level PPP projects benefited from more relaxed oversight. As the Organisation for Economic Co-operation and Development (OECD) rightfully noted in its report, the PPP preparation capacity is concentrated at the central level, although most of the activity takes place at the regional level, where there is much less capacity and expertise (Rivadeneira and Garin 2019). The PPP projects, therefore, created in the regions a commonly shared perception that the PPP model is a one-off and easy opportunity to win a stream of government-backed revenue lasting up to 30 years. It is no surprise, therefore, that many regions' officials and businesspeople with close ties to the regional and local authorities tried to ensure that a project was implemented through the PPP mechanism (often by way of direct negotiations and without any competition), rather than via the traditional procurement process under the Public Procurement Law that requires an open tender. Many of the PPP projects implemented in the regions during Stage Four, therefore, are associated with corruption—for example, the infamous corruption case in Mangistau region in 2019, where the region's deputy head has been sentenced to ten years in prison for "protecting" a foreign investor who proposed to construct an ice-rink stadium in Aktau by making use of the PPP model (Kapital 2019).

By the end of 2018, it became increasingly evident that PPP development requires significant adjustment and that the current PPP policy that enables exponential growth of the regional-level PPPs may wipe out Kazakhstan's state budget and seriously undermine the economy. The General Prosecutor's Office, the Ministry of Finance, and the National Chamber of Entrepreneurs (*Atameken*) each conducted their own investigations of the PPP deployment practice in the country and publicly confirmed numerous cases of fake PPPs, lack of transparency, and misuse of the PPP mechanism by the regional and local officials, including for corruption reasons (Akorda 2019). To address these issues, in May 2019, Kazakhstan's government announced a new PPP policy with project *quality* as a priority (in contrast to Stage Four policy, in which *quantity* of PPP projects was an unofficial priority).

Implementing the newly adopted PPP policy, the Ministry of the National Economy established a categorisation of all future PPP projects depending on the required compensation from the state budget: (i) category 1 projects require compensation of all expenses from the state budget (i.e. the "availability payment" PPP model), (ii) category 2 projects require compensation of the operating expenses only, and (iii) category 3 projects do not require any compensation from the state budget (i.e. the "users pay" model). Since 2019, the new PPP policy in Kazakhstan focuses, therefore, on giving preference to the "users pay" model for the projects (i.e. category 3) over the projects that require any financial commitment from the public sector partner (i.e. categories 1 and 2). To enshrine this new PPP policy in legislation, the Ministry of the National Economy also prepared a draft concept of law that, amongst other changes, would make the 2006 Concession Law applicable only to category 1 and category 2 projects (i.e. all large infrastructure projects that need government financial support would have to be implemented as concessions). The draft proposed that the 2015 PPP Law would be applicable only to category 3 projects that do not require any financial support from the public sector partner.

Policy Goals at This Stage

In addition to the new PPP policy goal of *quality instead of quantity*, the PPP Centre put forward an idea to use a new concept initiated by the United Nations Economic Commission for Europe (UNECE) called "People First PPPs" for development of the regional and local infrastructure projects in Kazakhstan. The main objective of this approach is to link PPP activities with public needs and involve domestic investors in providing public services and cultural activities (Rivadeneira and Garin 2019). In other words, it means that the People First PPPs approach requires that interests and needs of the local population be prioritised before the interests of the government (including political and national interests) and business (i.e. interest of the private firms to make profit). To put the People First PPPs concept in practice, the Ministry of the National Economy even introduced certain changes in its by-laws in December 2019 to ensure citizen engagement and improved budget management. At the regional and local levels, the approval for projects in the PPP form will be granted only to those projects that increase access of the local population, especially socially and economically vulnerable people, to essential services, or projects that promote social justice and increase the volume of public services.

Assessment

It remains to be seen whether a new PPP policy will be successful. However, it is already clear that this new policy has not been well developed. Neither the PPP Centre nor the government offered a clear understanding of what should be viewed as *quality* projects, as it seems that so far the only criterion for quality is the absence of requests for any financial support from the public sector partner. As for the People First PPP concept, it is likely to be just a declaration because its whole idea is self-contradictory. The PPP concept would secure the balance of interests of the government and business, and not the priority of interests of the local population, which would be a legitimate goal of certain socially oriented projects, rather than PPPs. Furthermore, the People First PPP concept would not offer sufficient help in solving the problems with PPP development encountered in Kazakhstan to date. The most tangible effect of the new PPP policy is that the number of executed PPP agreements in 2019 decreased by more than half compared to 2018.

Implications for Further Policy Development

There are concerns about whether the PPP Centre, which acts as a think tank for partnership promotion in Kazakhstan, proposed a proper new PPP strategy. The centre's official position is that Kazakhstan as a developing country does not need and cannot afford new large-scale infrastructure projects by means of PPP and, therefore, the strategy should be to implement small-scale and service-type (i.e. without capital investment) PPPs, such as kindergartens, public toilets, and clinics that do not require any financial support from the government (Yeshimova 2019). The centre's approach to PPP policy development appears rather narrow and is likely to jeopardise the attraction of foreign investment to the PPP schemes in Kazakhstan for a much-needed upgrade of, inter alia, the utilities and energy sector infrastructure.

Conclusion

Kazakhstan's substantial progress with developing PPPs and their enabling environment lags behind Russia only in the Eurasian Economic Union[2] (Eurasian Economic Commission 2018). However, there is still much room for improvement. As a financing mechanism, PPPs, if properly used, can be a useful tool for the government to not only attract private sector funds for upgrading its public infrastructure, but also to address corruption problems in public procurement and provide an effective legal framework for the ongoing privatisation of state-owned assets (e.g. by way of passing state-owned assets first under trust management of a private partner with a right to buy out a PPP facility at the end of the contract term).

In particular, Kazakhstan should tap into the huge potential of participating in China's One Belt and One Road Initiative by promoting PPP best practice to ensure high quality at lower costs. For example, the light rail system project in the city of Nur-Sultan, which was not structured as a PPP but rather with financing and investment coming from China, could have been implemented using the open tender in the PPP legal framework in the first place. The failure to structure this project as a PPP from the outset is now causing this project to fail and could become an embarrassment for the government, as well as lead to a wasteful use of the budget funds.

As the coronavirus outbreak spreads across the globe, it is likely that many existing PPP projects in Kazakhstan will experience considerable revenue generation challenges. At the same time, the need to attract private investment in Kazakhstan's healthcare system, agriculture, utilities, and other public infrastructure becomes even more acute. The time has come, therefore, to revisit the existing PPP policy. One option is to focus on the deployment of a larger number of adequately designed infrastructure PPPs at the national level with involvement of the IFIs and foreign

[2]The Eurasian Economic Union is an international organisation focusing on the regional economic integration. Currently, it includes the ex-Soviet countries of Armenia, Belarus, Kazakhstan, Kyrgyzstan, and Russia.

investors, while the number of local-level small-scale service-type PPPs should decrease.

References

ADB. (2017). *Public–private partnership monitor*. Available via http://dx.doi.org/10.22617/TCS179134-2. Accessed 4 April 2020.

Akorda. (2019, November 26). *Secretary of State conducted a meeting of the President's Commission on fighting corruption* [in Russian]. News item on the official website of the President of the Republic of Kazakhstan. Available via https://www.akorda.kz/ru/secretary_of_state/secretary_of_state_news/gosudarstvennyi-sekretar-provel-zasedanie-komissii-pri-prezidente-respubliki-kazahstan-po-voprosam-protivodeistviya-korrupcii?q=%D0%93%D0%A7%D0%9F. Accessed 26 April 2020.

Atameken. (2019). *PPP project financing: Issues and suggestions to address them. A presentation*. Available via https://forum.kzppp.kz/uploads/calendar/14/file3/3.%20Yakupbayeva%20Yulia.pdf?cache=1573484612. Accessed 4 April 2020.

Bassin, Y., Suleimenov, M., Osipov, E., & Chentsova, O. (2004). Subsoil use contracts: Issues of legal classification and systematisation. *Kazakhstan Business Magazine*, p. 1. Available via http://www.investkz.com/en/journals/38/166.html. Accessed 26 April 2020.

Beisembinova, A. (2011). Regulation of investment activity on the basis of public–private partnership [in Russian]. *Vestnik KazEU*. Available via https://articlekz.com/article/13982. Accessed 25 April 2020.

Boute, A. (2019). *Energy security along the New Silk Road: Energy law and geopolitics in Central Asia*. Cambridge: Cambridge University Press.

Chikanayev, S. (2015). Kazakhstan. In B. Werneck & M. Saadi (Eds.), *The public–private partnership law review* (pp. 144–165). London: Law Business Research Ltd.

Chikanayev, S. (2017). *A public–private partnership agreement as a new kind of an agreement in Kazakhstan's civil law and the problems of its categorisation* [in Russian]. Kaspiyskiy Obschestvenniy Universitet, Research publications series Adilet, No 3. Available via https://online.zakon.kz/Document/?doc_id=39697495#pos=36;70. Accessed 26 April 2020.

Eurasian Economic Commission. (2018). *PPP in countries of the Eurasian Economic Union*. Practical guide for investors [in Russian]. Available via http://www.eurasiancommission.org/ru/act/finpol/dobd/ppp/Documents/Evrazes_Book-2018_A4.pdf. Accessed 26 April 2020.

European Bank for Reconstruction and Development (EBRD). (2018). *2017/2018 Review: Kazakhstan*. Available via https://ppp-ebrd.com/review.php?country=Kazakhstan. Accessed 4 April 2020.

Kapital. (2019, August 27). *Ex-deputy head of Mangistau region was sentenced to 10 years in prison* [in Russian]. Available via https://kapital.kz/gosudarstvo/80767/eks-zamakima-mangistauskoy-oblasti-poluchil-10-let-lisheniya-svobody.html. Accessed 25 April 2020.

Kazakhstan Public–Private Partnership Centre. (2020). *PPP Centre*. Available via https://kzppp.kz/projects. Accessed 23 April 2020.

Law of the Republic of Kazakhstan. (2006). *Law on Concessions* [in Russian]. Law No 167 of 7 July 2006. Available via http://adilet.zan.kz/eng/docs/Z060000167. Accessed 26 April 2020.

Law of the Republic of Kazakhstan. (2015). *The Law on Public–private partnership* [in Russian]. Law No 379-V of 31 October 2015. Available via http://adilet.zan.kz/eng/docs/Z1500000379. Accessed 26 April 2020.

Matayev, T. (2014). *Concessions in the Republic of Kazakhstan* [in Russian]. Vestnik Finansovogo Universiteta, 3, 105–113. Available via https://cyberleninka.ru/article/n/kontsessii-v-respublike-kazahstan. Accessed 4 April 2020.

Moroz, S. (2006). *Investment law of the Republic of Kazakhstan* [in Russian]. Almaty: NII chastnogo prava KazGYuU.

Mouraviev, N., & Kakabadse, N. K. (2017). *Public–private partnerships: Policy and governance challenges facing Kazakhstan and Russia*. London: Palgrave Macmillan.

Mouraviev, N., Kakabadse, N., & Robinson, I. (2012). Concessionary nature of public–private partnerships in Russia and Kazakhstan: A critical review. *International Journal of Public Administration, 35*(6), 410–420.

Nazarbayev, N. (2012). *Strategy Kazakhstan-2050: New political course of the established state*. Available via https://strategy2050.kz/en/page/multilanguage/. Accessed 20 April 2020.

Newsworld. (2020, March 24). *SK E&C strikes financial contract with investors to finance Almaty Ring Road*. Available via http://www.newsworld.co.kr/detail.htm?no=6426. Accessed 4 April 2020.

Popondopulo, V. F., & Shevelyova, N. A. (Eds.). (2015). *Public–private partnership in Russia and foreign countries: The legal aspects* [in Russian]. Moscow: Infotropik Media.

Rivadeneira, A. M. R., & Garin, J. (2019) Public private partnerships review of Kazakhstan. *OECD Journal of Budgeting, 2*, 3–86. Available via https://doi.org/10.1787/f7696c94-en. Accessed 4 April 2020.

SZP. (2019). *Court cases about dispute resolution in the PPP projects in Kazakhstan* [in Russian]. Available via https://szp.kz/custom/wysiwyg/image/file/20191127/20191127115011_65888.pdf. Accessed 4 April 2020.

Tokayev, K.-J. (2019, July 15). *Speech of the Head of State at the meeting of the government* [in Russian]. Available via http://www.akorda.kz/ru/speeches/internal_political_affairs/in_speeches_and_addresses/vystuplenie-glavy-gos udarstva-k-tokaeva-na-rasshirennom-zasedanii-pravitelstva. Accessed 22 April 2020.

Tuktarov, Y., & Dubinchina, S. V. (Eds.). (2016). *A concession agreement: Its legal nature, scope of application, and execution* [in Russian]. Moscow: Statut.

UNECE PPP Assessment Report. (2013). *Green PPPs in Kazakhstan*. The United Nations Economic Commission for Europe. Available via https://www.unece.org/fileadmin/DAM/ceci/documents/UNDA_project/PPP_Readiness_Assessment_Kazakhstan.pdf. Accessed 26 April 2020.

Yescombe, E. R., & Farquharson, E. (2018). *Public–private partnerships for infrastructure: Principles of policy and finance*. Oxford: Butterworth-Heinemann.

Yeshimova, D. (2019). *Interview of the first Vice Chairman of Kazakhstan's PPP Centre to Khabar TV channel* [in Russian]. Available via https://www.youtube.com/watch?v=BMaOSOZRvUM. Accessed 4 April 2020.

CHAPTER 7

Water–Energy–Food Nexus Thinking in Kazakhstan: Choice or Necessity?

Marat Karatayev

Introduction

As three resources that are necessary for the human and natural system, water, energy, and food are increasingly closely linked. Food production requires a large amount of freshwater (about 70% of the total global freshwater withdrawals), and the generation of energy is inseparable from the use of water, accounting for 15–20% of water consumption (Food and Agriculture Organization [FAO] 2018). Similarly, food production and the food supply chain use about 30% of the total energy consumed globally, whereas energy represents 30–40% of production costs for farms, which includes cost of fertilisers, machinery, food transportation, and distribution, as well as extracting, pumping, lifting, collecting, and transporting water from aquifers in water-stressed areas (FAO 2018). All three

M. Karatayev (✉)
Ernst Mach Research Programme, Institute for Systems Science, Innovation and Sustainability Research, Karl-Franzens University of Graz, Graz, Austria
e-mail: Marat.Karatayev@uni-graz.at

A. Koulouri and N. Mouraviev (eds.),
Kazakhstan's Developmental Journey,
https://doi.org/10.1007/978-981-15-6899-2_7

sectors—water, energy, and food—are experiencing ever increasing pressure due to challenges such as population growth, urbanisation, and climate change.

It is estimated that by 2050 demand for food will increase by 70%, when the earth's population is projected to rise by 40% to over 9 billion people (Intergovernmental Panel on Climate Change [IPCC] 2007). At the same time, an estimated 1.3 billion tons of food are lost or wasted every year instead of being consumed (United Nations [UN] 2011). By 2025, in developing countries the total global water consumption will likely increase by 50% and in developed countries by 18%, while by 2050, over 40% of the world's population will live in severely water-deficient areas (UN 2014). Moreover, global primary energy consumption is projected to grow by 47% by 2035 (International Energy Agency [IEA] 2012) and water consumption for energy production will increase by 85%, driven by a shift towards higher-efficiency power plants that require more water and the increased production of biofuel—both leading to increasing water consumption (IEA 2012). Currently, about 90% of the energy production in the world is water-intensive, and 15% of the world's total water withdrawal is used for energy supply every year (World Resources International [WRI] 2016). In this context, the water–energy–food (WEF) nexus concept—treating each resource as a subsystem of a nexus, rather than a separate system (Endo et al. 2017)—has emerged as a useful innovative approach to address the complex and interrelated nature of resource systems in a sustainable manner at global, national, local, and basin levels.

The Need to Embed a Nexus Conceptualisation in Policymaking

Despite the WEF nexus concept being increasingly recognised as a useful framework for the governance of the three sectors by international organisations, academics, policy analysts, and other actors, decision-making and policy mechanisms are far from implementing this approach in practice (Al-Saidi and Elagib 2017). Different institutions governing the WEF nexus sectors and policy areas often do not communicate with each other and do not share information, resulting in a lack of integrated planning and management of WEF resources (Leck et al. 2015). For instance, key actors promoting food security often design agricultural and irrigation policies and subsidy programmes without considering the complexity of the inter-linkages with energy and water, and the impact on these resources, therefore missing opportunities to maximise the positive effects and minimise the damaging impact (Khan and Hanjra 2009). Similarly, in several countries, fossil fuel production, and renewable and nuclear energy programmes are designed without considering the consequences for degradation and depletion of water resources (Rivotti et al. 2019). Institutional processes need to recognise such implications and formulate and coordinate policies with appropriate legal, political, and administrative arrangements regarding suitable instruments to mitigate trade-offs and ensure high-quality WEF services while safeguarding the sustainability of the environment that supplies and maintains these services (Del Borghi et al. 2020).

Among the ways proposed for using the WEF nexus in practice is the cross-sectoral governance of the water, energy, and food sectors, based on relational equity theory. This approach involves formal and informal governing of the synergies and co-benefits between the different stakeholder groups, and also the collaborative management of the plurality and the interactivity of actions and policies that are in place (Koulouri and Mouraviev 2019a). Effective management of the relational equity of all parties seeks to promote understanding and facilitate the implementation of integrated governance. To achieve this, it takes into consideration the complexity of decision-making and the need for policy coherence in relation to the three nexus elements, while embracing political, social, financial, administrative, technological, and transboundary systems and promoting resource use efficiency, sustainability, and human development (Tankibayeva and Adibayeva 2019). Although policy and governance based on relational equity theory principles is important for decision-making to manage the interdependencies between the nexus elements, in practice cross-sectoral links in policymaking and implementation are often weak due to many horizontal and vertical challenges in coordinating the sectors. These challenges include an unequal distribution of power and varying competence across sectors, actors, and institutions; unclear rights and responsibilities of different stakeholders; lack of compatibility between institutional frameworks; sector-based policies and regulations; varying access to information and resources; lack of strategic clarity; and lack of political will (Rasul and Sharma 2016). Further challenges include conflicting interests within domestic policy, such as trade-offs between short-term gains and long-term sustainable solutions, and among actors within and beyond national borders (Rasul 2016). In order to lead to path-breaking change, the WEF nexus concept must therefore underpin and enable policy development and implementation.

Three Sectors: Current Conditions

Water Sector

Kazakhstan is one of the largest republics of the modern Commonwealth of Independent States (CIS).[1] Kazakhstan is located in the centre of the Eurasian continent and has an area of 2.7 million km^2. More than half (60%) of the territory of Kazakhstan is located in the desert and semi-desert zones with annual precipitation of less than 200 mm (Sinnott 1992). Due to climatic and geographic conditions, the country faces the problem of water scarcity. The total volume of Kazakhstan's renewable freshwater resources is on average about 100.6 km^3 per year, of which the volume formed in the country amounts to 55.94 km^3 per year with 44.64 km^3 per year flowing from the neighbouring countries, including China (19.2 km^3), Uzbekistan (14.7 km^3), Russia (7.6 km^3), and Kyrgyzstan (3.1 km^3) (FAO 2012). The water availability in Kazakhstan, on an annual basis, is 37,000 m^3 per km^2; the volume per capita is 3650 m^3, which is lower than the world average of around 6000 m^3 (Long et al. 2010).

Out of 167 countries studied, Kazakhstan ranked 20th out of the 33 countries that will experience extremely high water stress in 2040; that is, competition for and depletion of surface water (WRI 2019). The northern and south-eastern zones of Kazakhstan have the largest areas with extremely high water stress. This means that, in these regions, droughts are expected to occur on average every five years and the present hazard level may increase in the future due to the effects of climate change (Salnikov et al. 2015). Furthermore, water is unevenly distributed across the country, which presents a major challenge because the spatial patterns of water availability do not match the reality of demand for water and requirements for economic and social development. The Tobol-Torgai and Nura-Sarysu river basins in the central part of Kazakhstan account for 21% of the country's total population and approximately 35% of the total

[1] The Commonwealth of Independent States (CIS) is a regional intergovernmental organisation established in 1991. At present it includes Azerbaijan, Armenia, Belarus, Kazakhstan, Kyrgyzstan, Moldova, Russia, Tajikistan, and Uzbekistan.

arable land, but have only 3% of the total water resources in the country (Long et al. 2010). The Irtysh, Aral-Syrdarya, and Balkhash-Alakol river basins in the south-eastern part of the country account for almost 75% of the water resources generated within the country. The western part of Kazakhstan (the Ural-Caspian basin), the oil and gas province of the country, largely depends on groundwater and water desalination as a source of drinking water and water for irrigation, which uses around 30–35% of the total water supply, with the remaining 65–70% coming from surface water (UNDP 2004). Since the country gained its independence in 1991, water infrastructure planning, financing, provision, management, operation, and maintenance have deteriorated, reflecting the inefficiency of the water sector. One vivid indication of the poor functioning of the sector is that approximately 70% of the total water intended for irrigation does not reach the cultivated agricultural land (FAO 2012).

Energy Sector

Kazakhstan is rich in natural resources including coal, oil, natural gas, and uranium. Kazakhstan's total energy production is twice as high as its energy demand. This enables Kazakhstan to be a major energy exporter. In 2016, Kazakhstan was the 10th largest coal producer in the world. It is also among the top producers of crude oil (16th) and natural gas (23rd). Kazakhstan used a total of 78 million tonnes of oil equivalent (Mtoe) in 2016, while coal represented nearly half of Kazakhstan's energy mix (Kerimray et al. 2017; Energy Information Administration [EIA] 2017). In addition, coal accounted for more than 70% of the electricity production (72% of 106 terawatt-hour [TWh] in 2016), followed by natural gas (19%). Renewable energy including hydro energy accounted for 9% of the electricity generation (0.2 TWh from solar and wind, 9.3 TWh from hydro).

In 2018, the manufacturing sector accounted for approximately 70% of the total end use electricity consumption, with the residential sector accounting for 10%, the commercial and service sectors cumulatively for 9%, transport for 6%, and agriculture for 2% (Kazakhstan Electricity Grid Operating Company [KEGOC] 2018). Growth in the demand for electricity is forecasted to reach between 120 TWh and 180 TWh by 2030, that is equivalent to a 20–65% increase compared to the 2018 levels (KAZENERGY 2017).

In Kazakhstan, 80% of the total electricity is produced in the industrial north by power plants located near coal mines. The electricity transmission and distribution system is divided into three networks with two in the north connected to Russia, and one in the south connected to the Unified Energy System of Central Asia (Ahmad et al. 2017; Assembayeva et al. 2018). However, power transmission networks across the country are inefficient as loss of energy during transmission and distribution is estimated at 15%, although the actual percentage may be higher (KEGOC 2018).

Finally, in 2016, Kazakhstan's energy sector was responsible for carbon dioxide emissions of 275 $MtCO_2$ with 80% derived from heat and power generation owing to the low efficiency and ageing generating and network equipment (Kerimray et al. 2018).

Food Sector

About 40% of Kazakhstan's population is employed in agriculture and the sector contributes approximately 7% of the national gross domestic product (GDP) (FAO 2012). Between 1960 and 1988, Kazakhstan was the main wheat production area for the entire Soviet Union (Meng et al. 2000). However, after the country gained independence, the liberalisation of the agricultural sector led to a significant reduction in the output: between 1991 and 1997, it declined by 55%. Since 2000, agricultural production has shown a noticeable recovery and has exceeded the 1991 level (Broka et al. 2016). In 2016, Kazakhstan produced nearly 27 million tons of crops (Organisation for Economic Cooperation and Development [OECD] 2019). The UN Food and Agriculture Organisation estimated that the maximum crop production from Kazakhstan in good years exceeds 35–40 million tons (FAO 2012). Wheat is the main crop in the country, and 14–16 million tons of grain are produced annually. Around 75–80% of Kazakhstan's wheat production is in the north, including the Tobol-Torgai and Ishim river basins, with a small amount being produced in the south (FAO 2017). The east and south of the country are areas of fibre and non-grain crop production, including cotton, flax, sugar beet, and tobacco. The south, east, and central Kazakhstan regions are also major areas of fodder crop production, including barley, oilseed rape, grass, maize, millet, soya beans, and oats.

One of the major concerns for sustainable agricultural production is the continued soil erosion and degradation due to salinisation, acidification, and contamination by anthropogenic pollution. According to the Kazakhstani Research Institute of Geography (KIG 2013), most of the country's territory (179.9 out of 272.5 million hectares—66% of the total area) has a moderate level of soil degradation; 31.3% of total irrigated lands are salinised; and all degraded arable lands have lost soil organic carbon humus (KIG 2013). There are areas with high levels of soil degradation across the western, northern, and south-eastern zones of Kazakhstan, which challenges agricultural productivity.

Intersectoral Usage of Nexus Resources

Water-Related Energy Usage

Kazakhstan's total water-related energy consumption is 14.50 TWh; the energy-related cost of every unit of water used in Kazakhstan is estimated at 0.65 kilowatt-hour (kWh) per m^3 (KIG 2013). The most costly stages of the Kazakhstani water use cycle are the extraction and water treatment stages, which account for 60% of the total water-related electricity demand. Irrigated agriculture in Kazakhstan is poorly managed and now accounts for 45% of Kazakhstan's total water-related electricity demand (Rivotti et al. 2019). Wastewater treatment accounts for 16% of the water-related electricity demand. Water treatment has the broadest range of energy intensity in the water use cycle because water treatment depends entirely on the quality of the source water. In Kazakhstan, as in most countries suffering from hydrological stress, desalination has become an alternative water source particularly in western parts of the country, which explains the increasing energy demand for water treatment.

Energy-Related Water Usage

The energy sector accounts for only 7.5% of the total water-related energy usage, but in terms of extracted volume, it is responsible for 25% of the total water withdrawn. This volume of withdrawn water includes water consumption for the extraction and refining stages, and for power generation. The total water demand for traditional oil and coal extraction is 155.05 million m^3, while 114.51 million m^3 is used for thermal electric cooling (WRI 2019).

It is expected that future water consumption will increase as Kazakhstan transforms its energy sector. By 2040, onshore oil and coal extraction are projected to decrease by 12 and 20%, respectively, while uranium production is likely to remain the same as in 2014. In the petroleum-refining sector, the three main oil refineries in Kazakhstan are assumed to continue operating through to 2040, with a capacity increase of 20%. In the power generation sector, the total thermal generation capacity fuelled by resourcesfossil is projected to remain approximately the same as in 2014. However, the share of gas-fuelled electricity generation is expected to increase by 31%, while coal usage for electricity generation will decrease by 50%. This assumption is predicated on Kazakhstan's government's ambitious plans to significantly reduce carbon emissions per unit of GDP by 25% in 2050 (Wang et al. 2019). It is expected that the additional capacity required to meet future demand will be provided by the introduction of 1.2 gigawatt (GW) of nuclear generation and a small share of wind energy (the "business as usual" scenario for 2040 includes 3 GW of new wind capacity). As a result of this energy transition, future nuclear power will account for 50% of Kazakhstan's water withdrawals by the energy sector, whereas coal-based thermal power production will account for 32% (Rivotti et al. 2019).

Water Footprint for Biofuels

Given the current and projected coal use in Kazakhstan and the need to cut emissions, producing biofuels and applying biomass co-firing in coal power stations is viewed as a promising option that could enable the country to decrease its carbon emissions and provide an improved sustainable energy supply. The theoretical bioenergy potential in Kazakhstan is estimated at 485.36 megajoules (MJ) or 16.582 million tonnes of coal equivalent, with an average of 14.150 metric tonne per year (tonne/year) in the past 12 years (Koshim et al. 2018). This is equivalent to around 30% of the total current energy consumption in the country.

In Kazakhstan, biomass energy resource mainly consists of wheat residues (44%), produced in the northern part of the country—in the Tobol-Torgai and the Ishim river basins. The biomass energy potential from wheat residues is found to be 3612.0 tonnes/year in the Tobol-Torgai river basin, 3101.4 tonnes/year in the Ishim river basin, and 2846.3 tonnes/year in the Irtysh river basin (Koshim et al. 2018). The southern part of Kazakhstan (Aral-Syrdarya and Chu-Talas-Assa river basins) also shows potential with a maximum of 2105.5 tonnes/year of biomass residues.

However, the development of the bioenergy sector in Kazakhstan will have consequences owing to projected increase in water and land usage. Based on the stated assumptions, the amount of water required for agriculture would increase by 20% for dry land agriculture (i.e. if biomass is grown on dry land) and by 35% for irrigated land agriculture (i.e. if biomass is grown on irrigated land). Considering the problems that basin agencies face with providing the current volume of water required for agriculture, this increase in water usage is clearly a non-viable alternative for the country. Providing the required supplementary land area is also worrisome: supplementary dry land area would account for 20–25% of the total agricultural area and supplementary irrigated land area would account for 15%. For an arid country like Kazakhstan, and its generally low yields under rain-fed regimes, such land conversions would severely impact on agricultural production.

Water and Energy Use in Agriculture

The current water intake for agriculture is 13.4 km^3 per year (69.1%), of which 8.8 km^3 per year (65.6%) is lost during transportation along canals, with only 3.8 km^3 per year being used for regular irrigation purposes (FAO 2017). Water use efficiency in Kazakhstan is significantly lower than in industrialised countries. According to the FAO data, about 3500 m^3 of water was used to produce 1 ton of crops in Kazakhstan. It took 1300 m^3 of water to produce the same volume of crops in Poland, 1000 m^3 in the United States, 790 m^3 in the UK, and 660 m^3 in France (FAO 2012). At the same time, the geophysical and climatic conditions of the country present a serious challenge to the development of its agriculture sector. Due to the location of cultivated land in the desert and semi-desert zones, 60% of agricultural land is irrigated, which is the highest percentage in Central Asia. Adequate and uninterrupted energy supply is required for the proper functioning of irrigation systems and the supply of water. Apart from that, diesel and, to a lesser extent, gasoline are needed to power farm machinery, while large volumes of natural gas are required for producing fertilisers and pesticides. In 2016, the agricultural sector in Kazakhstan consumed 6.4 Mtoe of direct and indirect energy, accounting for about 15% of the total national primary energy consumption. Electricity use in this sector accounts for 747 GWh (KEGOC 2018).

Regional Problems in Utilisation of the Nexus Resources

The Irtysh, Aral-Syrdarya, and Balkhash-Alakol river basins are rich in water resources, and a large part of their population has access to water and electricity. The Irtysh river basis is the most secure in terms of availability of its water resources; the water reserve comprises 43.8 km^3 (FAO 2017) with the principal water reserves (60%) formed by the river flow—26.04 km^3. However, the Irtysh basin has the highest level of energy consumption in relation to the GDP, undermining future energy security of this region. In the Balkhash-Alakol basin, the water reserve is significant, amounting to 149.4 km^3. However, the main volume of water is concentrated in the mountains, mainly in the Balkhash Lake, and cannot be used for irrigation within most of territory of south Kazakhstan (FAO 2017). In terms of energy security, the Balkhash-Alakol basin is the second largest energy consumer as a large part of electricity is received from energy-intensive coal-based power stations. The Nura-Sarysu and Tobol-Torgai river basins are the largest consumers of the nexus resources. These basins also exert significant environmental pressure by contaminating water, air, and land resources. The specific feature of the rivers in these basins is that the main annual water discharge (90% or more) happens during the short spring flood. During low water periods in the summer, autumn, and winter, the discharge of the rivers significantly decreases, and in most rivers no discharge is observed. The Nura-Sarysu and the Tobol-Torgai river basins use nexus resources outside of each basin, which account for approximately 60% of the water, 25% of the energy, and 75% of the agricultural resources required (Broka et al. 2016). The Nura-Sarysu river basin's priorities place less importance to the agriculture sector (due to soil degradation and salinisation) and emphasise the development of the heavy industry. Promoting energy efficiency technologies in the heavy industry is crucial for energy security in these river basins.

External Factors Affecting the WEF Nexus

One of the significant external factors affecting the WEF nexus in Kazakhstan is the transboundary cooperation among the nations of the Central Asian region. Before the disintegration of the Soviet Union in 1991, the regional energy network was governed by the Unified Energy System of Central Asia that included all five Central Asian Soviet republics. In this network, electricity was freely traded and supplied throughout the whole region. The system was created to mitigate the region's disproportionate distribution of natural resources. The upper countries of the region (Tajikistan and Kyrgyzstan) have large water reserves and almost no hydrocarbon resources, whereas the downstream countries (Uzbekistan, Kazakhstan, and Turkmenistan) have large oil, gas, and coal reserves but suffer from water shortages (Abdolvand et al. 2015). During the summer, water-rich upstream countries were responsible for supplying water for irrigation to downstream countries to support their agriculture, whereas during the winter, energy-rich countries provided fossil fuels and supplied electricity to Tajikistan and Kyrgyzstan. This system was crucial for Kazakhstan, as more than 60% of water demand for the needs of the republic's agriculture was covered by water supply from other Central Asian Soviet republics. However, after the collapse of the Soviet Union, this system gradually fell apart, as international disputes arose. Each of the five newly independent states focused on its national interests and aimed to ensure its own national energy independence. Since 1995, Kazakhstan and Uzbekistan have stopped supplying energy resources and electricity to Kyrgyzstan and Tajikistan. This allowed these two water-rich countries (Kyrgyzstan and Tajikistan) to release large volumes of water (that could be used for irrigation) in winter, resulting in Kazakhstan facing frequent shortages of water for agriculture in summer.

Policy Response to the WEF Sectors' Challenges

To respond to the interrelated difficulties faced by the nexus sectors within the context of national challenges, such as a need to ensure economic growth, increasing population, environmental pollution, and resource scarcity, Kazakhstan has introduced various policy and regulatory measures (Fig. 7.1), set short- and long-term objectives and designed the governance structure by assigning different institutions responsibility for the WEF sectors (Table 7.2). This section provides a brief overview of the policy initiatives.

Under its obligations according to the Kyoto Protocol that Kazakhstan signed in 1999 (Presidential Decree 1999) and the Paris Agreement on Climate Change ratified by Kazakhstan in 2016 (Law of the Republic of Kazakhstan 2016a), the country voluntarily took obligations to achieve

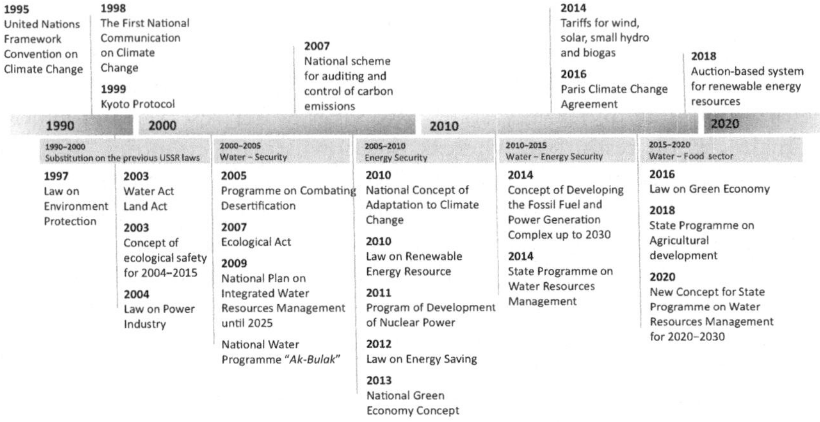

Fig. 7.1 Timeline of introduction of key WEF nexus policies

the following quantitative targets: by 2030, reduce greenhouse gas emissions by 15% and by 2050, reduce by 40% in comparison with the 2012 emission level (Table 7.1).

Regarding the nation's *energy* needs, currently, Kazakhstan has an annual population growth rate of about 1.3% with the present population of 18.2 million being projected to reach 26 million by 2030 (World Bank 2020). Furthermore, the country has experienced economic growth with an average annual rate of about 3–4%. In this context, the domestic energy generation and use have been growing rapidly. Between 1999 and 2015, the primary energy consumption grew from 26.92 Mtoe to 91.08 Mtoe. It is expected that by 2030 the primary energy production will be between 140 Mtoe and 160 Mtoe. In 2014, the Ministry of Industry and New Technologies put forward the Concept of Developing the Fossil Fuel and Power Generation Complex until 2030 (Concept of Developing the Fossil Fuel and Power Generation Complex in Kazakhstan until 2030 2014). The concept envisages the increase in the volume of coal mining from 103.5 million tonnes per year, in 2014, to 113 million tonnes by 2030. Oil production is projected to increase from 84.2 million tonnes per year, in 2014, to 118.1 million tonnes by 2030, and natural gas production from 44.2 billion m^3 per year to 59.7 billion m^3. In addition, by 2050, the government intends to increase the share of gas in the national energy mix by 25% compared to the 2015 level. The sectoral programme for

Table 7.1 Policy documents and objectives set for WEF nexus sectors

Policy document	*Objectives*	*Indicators*	*Sector*
Concept of Ecological Safety for 2004–2015 (approved on 3 December 2003). The concept became null and void by the President's decree of 13 April 2011	• Prevention of depletion and pollution of water resources • Solution of transboundary water problems • Reduction of anthropogenic pressure, leading to climate change and destruction of the ozone layer	15% infrastructure upgrade	Water
National Plan for Integrated Water Resources Management for the period 2009–2025 (approved on 28 January 2009) (Government Decree 2009)	• Adoption of water river basin approach and reduction of water lost in agriculture	30% by 2025 8 River Basin Councils based on hydrological boundaries	Water
Water Programme *Ak Bulak* for the period 2011–2020 (approved on 9 November 2010) (Government Decree 2010)	• Providing reliable access to water resources	100% by 2020 in urban areas 80% by 2040 in rural areas	Water
Law on the Utilisation of Nuclear Energy (approved on 12 January 2016) (Law of the Republic of Kazakhstan 2016b)	• Increasing the share of nuclear power in the national energy mix	Nuclear power plant with a capacity from 900 to 2000 MW by 2030	Energy
The concept of Kazakhstan's transition to a "green economy" (approved on 30 May 2013)	• Reduction of GDP energy intensity	25% by 2020 30% by 2030 50% by 2050	Energy

(continued)

Table 7.1 (continued)

Policy document	*Objectives*	*Indicators*	*Sector*
	• Development of RES through the construction of wind, solar, biomass and small hydro power facilities	3% by 2030 50% by 2050	Energy
	• Reduction of current CO_2 emissions in electricity production	Levels of 2012 −15% by 2030 −40% by 2050	Energy
Concept of Developing the Fossil Fuel and Power Generation Complex in Kazakhstan until 2030 (approved on 28 June 2014)	• The increase in volume of fossil fuel production for national energy security	Coal: 113 tonnes per year by 2030 Oil: 118.1 tonne per year by 2030 Gas: 59.7 billion m^3 per year by 2030	Energy
State Programme for the Development of Agriculture in Kazakhstan for 2017–2021 (approved on 12 July 2018)	• Reduction of irrigation water consumption	Level of 2017 −15% by 2021	Water
	• Increasing wheat productivity	Level of 2017 +60% by 2030	Food
	• Reduction in yield and price volatility	25% of areas with efficient technologies	Food
The concept of water resources management for 2020–2030 (approved on 28 January 2020)	• Reduction of water consumption per unit of GDP	Level of 2018 −20% by 2030	Water
	• Increasing the irrigated agriculture land	43% by 2030	Food
	• Increasing the length of lined water pipeline	from 3423 to 19,000 km	Water

Source Compiled by the author

Table 7.2 WEF nexus governance organisations

Government ministry	*Departments*	*Nexus sectors*	*Government policy documents*
Ministry of Agriculture	• Committee on Water Resources • Committee on Land Resources • Department of Production and Processing Livestock Products • Department of Production and Processing Plant Products • State Inspection Committee in the Agro-industrial complex	Water Food	• Water Code (2003) • Land Code (2003) • Water Programme *Ak Bulak* for 2011–2020 (Government Decree 2010) • State Programme for the Development of Agriculture in Kazakhstan for 2017–2021 (2018)
Ministry of Energy	• Committee for Atomic Energy Supervision and Control • Department of Oil and Gas Industry • Department of Nuclear energy and Industry • Department of Renewable Energy Development • Department of Electric Power Development • Department of Climate Policy and Green Technology	Energy	• Law on the Electrical Power Industry (Law of the Republic of Kazakhstan 2004) • Ecological Code (2007) • Law on the Utilisation of Renewable Energy Sources (Law of the Republic of Kazakhstan 2009) • The concept of Kazakhstan's transition to a 'green economy' (2013)

(continued)

Table 7.2 (continued)

Government ministry	*Departments*	*Nexus sectors*	*Government policy documents*
Ministry of Ecology, Geology, and Natural Resources	• Department of Transboundary Rivers Management • Department of Environmental Policy and Sustainable Development • Department of State Policy in Waste Management • Department of Geology and Natural Resources Management	Water Energy	• National Plan for Integrated Water Resources Management for the period 2009–2025 (Government Decree 2009) • Concept of Developing the Fossil Fuel and Power Generation Complex in Kazakhstan until 2030 (2014) • The Concept of Water Resources Management for 2020–2030 (2020)

Source Compiled by the author

utilisation of nuclear power in Kazakhstan was also developed (Law of the Republic of Kazakhstan 2016b). As for renewable energy, the immediate target is to increase its share to 3% of the total energy mix by 2020 and to 6% by 2025. By 2050, there is a much more ambitious target of 50% of the total power to be generated from renewable energy sources (Presidential Decree 2013).

In order to achieve these targets, the government proposed a number of tools and incentives (Fig. 7.1 and Table 7.1). The country approved tariffs for energy produced from wind, sun, small hydro, and biogas facilities, and since 2018 an auction-based scheme has been operating in which different projects compete for financial support. A certain amount of electric capacity is tendered, corresponding to the national energy goals of generating electricity from renewable energy sources (Boute 2020), which was envisaged by the Roadmap for the Development of Alternative Energy for the period 2012–2030. According to the Roadmap, the plan is to launch 106 renewable energy projects with more than 3000 MW of renewable energy capacity by 2020. In addition, different electrification programmes using energy from small hydropower plants are currently running. The government intends to construct 11 small-scale hydropower stations with 205 MW by 2020 in Almaty, and eastern and southern Kazakhstan regions. To support the transition to a sustainable energy system, several laws to address the energy challenges have been adopted, including the Law on the Electrical Power Industry (Law of the Republic of Kazakhstan 2004), Law on the Utilisation of Renewable Energy Sources (Law of the Republic of Kazakhstan 2009), and Law on Energy Saving and Increasing Efficiency of Energy Utilisation (Law of the Republic of Kazakhstan 2012).

With respect to *water* policy, a step forward has been made when the country approved in 2020 a new concept for the state programme on water resources management for 2020–2030 proposed by Kazakhstan's Committee on Water Resources (Zona KZ 2020). The new concept focuses on key areas such as international cooperation, updating the legal framework for water use and management, institutional reform, modernisation of the water industry infrastructure, making use of the international best practice in creating water market, digitalisation of the water sector, the water utilisation in environmentally friendly way, training of experts for the water industry, and implementation of national water projects. It is expected that all these dimensions of water management will meet the population's needs, and preserve and restore the country's water resources. The new concept also involves developing collaboration between the public and private sectors, such as reimbursing part of the costs of construction or modernisation of facilities, transferring public property to private individuals in trust or lease, and co-financing leasing projects. In addition, the new concept aims to maintain the water balance at a level of 100 km^3 by 2030 by increasing the surface water resources through the construction of new water reservoirs (an increase of 5–7 km^3), reducing water consumption (savings up to 5 km^3), and increasing groundwater use (increase up to a total of 15 km^3). There are additional plans to reduce water consumption from 91.2 m^3 to 73.0 per USD 1000 of the GDP, build 26 new hydropower facilities, and reconstruct 182 national and 300 municipal water facilities. The programme provides for the construction of new irrigation systems to increase the arable agricultural land from 1.7 million to 3 million hectares, and increase the length of water pipeline from 3423 to 19,000 km.

Kazakhstan is not a *food* deficient country, compared to other Central Asian states. Furthermore, Kazakhstan is now the main grain exporter in the region. However, due to inefficient agricultural practice and management problems there is significant environmental impact. If it is intensified by climate change and a decrease in water availability, it could lead to the destabilisation of the agricultural sector, posing a threat to food security and Kazakhstan's agricultural development. There is a sectoral programme for agricultural development (Government Decree 2018), aimed at providing both national food security and increasing agricultural export. According to the programme, the diversification of the crop production will take place as a result of optimising crop areas. Although there is a plan to reduce the area under grain cultivation from 17.2 million hectares to 15.2 million hectares by 2021, the area for durum wheat will triple to further develop exports. In addition, productivity is expected to double, agricultural export will reach 8% of the country's total export, and domestic food products will satisfy more than 80% of the internal market's requirements. To support agricultural development, the programme on combating desertification was launched in 2005, although it was stopped a few years later Government Decree (2005).

Governance of the WEF Nexus: What Are the Issues?

The Ministry of Agriculture, the Ministry of Energy, and the Ministry of Ecology, Geology, and Natural Resources are the key actors involved in the governance of the WEF sectors. In addition, these ministries have organisational structures at subnational levels of the government. The Ministry of Agriculture is a central national authority on food security, whereas responsibility for water management is divided between the Ministry of Agriculture and the Ministry of Natural Resources. The Ministry of Natural Resources and the Ministry of Energy are jointly responsible for the implementation of energy policy.

A number of policy initiatives targeting the WEF sectors in Kazakhstan have been developed by these ministries (see Fig. 7.1). However, responses to the challenges the three sectors face are generally narrowly focused, and actions are taken in isolation, without consideration of their impact on other sectors (Koulouri and Mouraviev 2019b). The importance of adopting the WEF nexus approach to ensure rational resource utilisation is still not recognised when sector-focused experts draft government policy documents and related legislation, or design national programmes (Janusz-Pawletta 2015). There is a lack of effective coordination, on the one hand, and duplication of activities and responsibilities within and between various agencies involved in management of the WEF sectors at the national level, on the other hand (O'Hara et al. 2008). This is due to the fact that the current legislative framework does not adequately clarify the functions of different governance institutions regarding intersectoral kinds of problems, and also regarding issues that arise between the central, regional, and local governments.

The situation is further exacerbated by the existing governmental structures and governance processes. Kazakhstan's governance structure includes three levels: central (national), regional (*oblast*) and local. Local authorities (responsible for the governance of the WEF sectors) continue to serve as the instruments of the central and regional governments. At the same time, the local governments are legally established as independent authorities; however, in practice, their authority is weak. As studies show, in the case of the water sector governance (Janusz-Pawletta 2015) and the development of renewable energy projects (Karatayev et al. 2016), the relationships between local authorities are unclear and their interaction is almost non-existent. Similarly, the relationships between local authorities and central government agencies are unclear. Local authorities have limited influence on institutional and legislative development, and most of their opportunities at a practical level are limited. In addition, the central government does not provide opportunities for them to contribute to political debates or policy development. Furthermore, budgets of local authorities are dependent on allocations from the central government (via regional governments) as local authorities have the right to collect only a limited range of local fees and taxes. Control of public resources by the central government results in a relationship of dependency, which reinforces the upward accountability structures and limits the autonomy of the local authorities. The lack of autonomy and authority to influence political decisions leaves local governments unable to manage the WEF sectors adequately and in a sustainable manner, as they remain under the tight control of the national ministries and are guided by national policies and the national regulatory and legislative frameworks, rather than by local priorities.

Within the *energy sector*, most of the policy objectives included in the concept of Kazakhstan's transition to a green economy are related to carbon emissions, while no water policy objectives have been set (see Table 7.1). The green economy concept developed by the Ministry of Energy does not include water-related objectives, water availability aspects, or conditions for hydropower generation. Similarly, the programme for the development of agriculture for 2017–2021 was designed without consulting water stakeholders and without considering water availability scenarios. For ensuring sustainability, the green economy concept does not include indicators related to water, such as water volume for cooling thermal power stations, integration of energy infrastructure with the water network system, the use of land areas for the production of hydropower, the use of fishing areas, and the occupation of areas subject to risk of flood. The impact of the energy sector on water availability and quality is largely disregarded (Rivotti et al. 2019).

Furthermore, numerous agencies are involved in the energy sector governance, which provides an additional example of the departmentalisation and fragmentation of governance in the WEF sectors: the focus is on certain aspects and issues of the sector while its holistic consideration is lacking. Specifically, the role of the Ministry of Energy is to create conditions for the development of the energy market and ensure adequate energy supply. The ministry is responsible for all activities related to the exploration and development of minerals and hydrocarbons. However, as for economic regulation, the ministry's role is limited. The energy market is regulated by the Agency for Regulation of Natural Monopolies (ARNM). The agency has the privilege to set the electricity tariffs and charges for services related to the sale of electricity. The agency's approval is required if an energy provider intends to raise a tariff. The ARNM also awards licences to power providers and distributors, and monitors compliance with the terms of the licences. The Kazakhstani Electrical Grid Operating Company (KEGOC) has been responsible since 1991 for the transmission of electrical energy throughout Kazakhstan, from the power stations to the companies that sell electricity to end-users (Karatayev et al. 2016).

A similar situation can be observed in the governance of the *water sector* in Kazakhstan. In 2003, Kazakhstan adopted a national water code (Water Code of the Republic of Kazakhstan 2003) that increased the recognition of the importance of water management at the local level through adopting the river basin approach, while the ultimate decision-making power remains centralised and is vested with the different ministries. In particular, the Ministry of Agriculture is responsible for water allocation and its use for irrigation in the country and has the Committee on Water Resources charged with working on the water sector issues. The committee also provides oversight and policy advice to the River Basin Councils (RBC). The Ministry of Industry and Infrastructure Development manages drinking water and wastewater, the Ministry of Health oversees water quality, and the Ministry of Energy regulates water contamination in relation to extractive industries. At the local level, allocation of water rights and resources is managed by water user associations and village irrigation committees on a communal basis and according to customary rules and practices. In addition, there are water institutions, regional organisations, and private companies (e.g. Kazagroexport and the Water and Agriculture Centre, *Kazagromeliovodhoz*) that have a role in decision-making affecting the country's water policy. Each of these organisations has its own mission, legal authority, and responsibilities. Many operate as independent or quasi-independent organisations. Some are government agencies, and others are incorporated as non-profit organisations with their own internal governance structures (Janusz-Pawletta 2015).

Experts in agriculture and food security note that the operation of the *food sector* in Kazakhstan is plagued with large-scale ineffectiveness (Rhoe et al. 2008). The root cause of this ineffectiveness is the centralised, top-down, state control, while central government institutions are reluctant to devolve responsibilities to the regions and localities. In addition, as in the case of the water and energy sectors, the processes and frameworks to support the interaction between various actors within the sector and beyond the sector's boundaries are poorly designed (Baydildina et al. 2000). Much like in the energy and water sectors, there are many different kinds of organisations involved in the governance of the food sector. The institutional actions on water issues in agriculture are the responsibility of the Ministry of Natural Resources, while their implementation is mainly the prerogative of the Ministry of Agriculture. Water planning is made by the Ministry of Natural Resources, while decisions on what types of crops to grow, what types of irrigation technologies to use, and how to power those technologies are taken by the Ministry of Agriculture at the national level. Moreover, the Ministry of Agriculture is mandated to promote, monitor, and regulate the agricultural sector. However, the Ministry has no legal jurisdiction over land allocation to different uses or the management of the irrigated areas, except over issues related to combating diseases and conducting chemical analysis for vegetables entering the market. Despite certain constraints related to its governance capacity, the ministry designed and implemented national policies, programmes, and projects (e.g. State Programme for the Development of Agriculture in Kazakhstan for 2017–2021 [Government Decree 2018]) aimed at achieving agricultural growth through better utilisation of the country's soil and plant resources, although water and energy factors were largely not taken into account.

Conclusion: Towards the WEF Nexus Governance

The WEF nexus concept is a new phenomenon for Kazakhstan. The usefulness of the nexus approach was acknowledged across the globe relatively recently, in 2011.[2] Some successful WEF nexus governance practice

[2] The WEF nexus approach gained momentum after the Water, Energy and Food Nexus Conference in Bonn in 2011 (Hoff 2011) and the Annual World Economic Forum, which brought the Water, Energy, and Food Security Nexus to full political attention at the Davos Summit (2011) through the Global Risks 2011 report (World Economic Forum 2011).

could be learnt from European countries (e.g. see Howarth and Monasterolo 2016). Kazakhstan is currently facing numerous interconnected challenges, including ensuring water availability and a reliable supply of energy and food at both national and local levels. At the same time, the current governance system of the WEF sectors in Kazakhstan is still departmentalised, fragmented, and bureaucratic and gives the providers limited flexibility, while their accountability is far from perfect. It is therefore unlikely to ensure the sustainable and efficient management of WEF resources. The adoption of a WEF nexus approach appears necessary to address the complex, inter- and cross-sectoral challenges faced by the three sectors and to support cross-sectoral collaboration and integrated development planning required to meet these challenges.

A WEF nexus approach will bring a number of advantages for the three sectors, as it could improve resource use efficiency, increase the reliability of supply, and strengthen governance cohesion to achieve sustainable development. The WEF nexus approach could be viewed as a framework that allows to connect actors, coordinate planning and implementation, and monitor management for the three sectors by paying special attention to identifying and conceptualising interconnections and trade-offs, in order to build aligned and coherent strategies and actions that permit synergies. The nexus framework is applicable at the national, regional, and/or local levels, and on an enterprise or even a household scale, where resource management and optimisation are also required. The framework's application could be underpinned by the concept of relational equity management (Koulouri and Mouraviev 2019a), including in the governance processes many diverse actors at various levels and scales, for example, a manufacturing enterprise, small-holder farmer, and individual water consumer.

For example, in the water sector, a WEF nexus approach can eliminate the focus on food and agriculture by formally recognising a greater plurality of water users, including the energy sector. Furthermore, it can change the role of River Basin Councils and establish them as new institutions, which will allow the energy sector (that never previously participated in the river basin's planning) to become involved in, and influence, water resource management .

Importantly, creating conditions under which nexus thinking will become a norm in the nation's policymaking and implementation is required, including changes in the thinking of the central government and its executive agencies. The expertise and competence of other public and private actors—at all levels, from national to local—would have to be more widely recognised and valued. Efficient resource use policies cannot be drafted without collaborative work of many, rather than a few, actors. Furthermore, it has to be acknowledged that the change requires not only consultation with a broader range of stakeholders, but a call for the central/national organisations (e.g. ministries) to relinquish some of their authority and reposition themselves as facilitating and enabling organisations, rather than policymaking bodies.

Governance mechanisms at the local level, involving a wider range of participants, could provide valuable opportunities for widening the knowledge base and legitimising decisions made by the upper levels of government. Nonetheless, despite certain benefits from actors' engagement in the nexus governance at the local level, the adoption of a collaborative approach in an environment characterised by a highly centralised nation's public administration would be a challenging departure from existing norms and practice. The importance of the WEF nexus approach and the need to understand the trade-offs, side effects, and synergies related to collaborative nexus governance are yet to be fully embedded in the nation's prevailing and emergent policy paradigms.

Acknowledgements The author would like to express his sincere thanks and gratitude to the Karl-Franzens University of Graz, in particular the Director of the Institute of Systems Sciences, Innovation and Sustainability Research, Professor Rupert Baumgartner, and Associate Professor Thomas Brudermann. For financial support, the author would like to thank the Austrian Agency for International Cooperation in Education and Research (OeAD) for the award of the Ernst Mach Grant (IND130547) "Development of a methodology for measuring energy security and risks for balancing energy security and risk for resource-rich countries in the climate constrained world". The author also gratefully acknowledges the Ministry of Education and Science of the Republic of Kazakhstan for the national grant (AP05135081) "The problem of energy security within the Eurasian Economic Union' (EAEU) energy policy-making framework", coordinated by Professor Galiya Movkebayeva (Al-Farabi Kazakh National University).

References

Abdolvand, B., Mez, L., Winter, K., Mirsaeedi-Gloßner, S., Schütt, B., Rost, K. T., et al. (2015). The dimension of water in Central Asia: Security concerns and the long road of capacity building. *Environmental Earth Sciences, 73*(2), 897–912.

Ahmad, S., Nadeem, A., Akhanova, G., Houghton, T., & Muhammad-Sukki, F. (2017). Multi-criteria evaluation of renewable and nuclear resources for electricity generation in Kazakhstan. *Energy, 141,* 1880–1891.

Al-Saidi, M., & Elagib, N. A. (2017). Towards understanding the integrative approach of the water, energy and food nexus. *Science of the Total Environment, 574,* 1131–1139.

Assembayeva, M., Egerer, J., Mendelevitch, R., & Zhakiyev, N. (2018). A spatial electricity market model for the power system: The Kazakhstan case study. *Energy, 149,* 762–778.

Baydildina, A., Akshinbay, A., Bayetova, M., Mkrytichyan, L., Haliepesova, A., & Ataev, D. (2000). Agricultural policy reforms and food security in Kazakhstan and Turkmenistan. *Food Policy, 25*(6), 733–747.

Boute, A. (2020). Regulatory stability and renewable energy investment: The case of Kazakhstan. *Renewable and Sustainable Energy Reviews, 121,* 1–11.

Broka, S., Giertz, Å., Christensen, G. N., Rasmussen, D. L., Morgounov, A., Fileccia, T., et al. (2016). *Kazakhstan—Agricultural sector risk assessment* (Agriculture Global Practice Technical Assistance Paper). World Bank Group Report. Available via http://documents.worldbank.org/curated/en/422491467991944802/pdf/103076-WP-KZ-P154004-Box394863B-PUBLIC-ASRA.pdf. Accessed 3 March 2020.

Concept of Developing the Fossil Fuel and Power Generation Complex in Kazakhstan until 2030. (2014). *Decree No 724 of 28 June* [in Russian]. Available via https://online.zakon.kz/Document/?doc_id=31581132. Accessed 9 January 2020.

Concept of Ecological Safety for 2004–2015. (2013). *Decree No 1241 of 3 December* [in Russian]. Available via https://online.zakon.kz/document/?doc_id=1045395. Accessed 11 April 2020.

Del Borghi, A., Moreschi, L., & Gallo, M. (2020). Circular economy approach to reduce water-energy-food nexus. *Current Opinion in Environmental Science & Health, 13,* 23–28.

Ecological Code. (2007). *Ecological Code of the Republic of Kazakhstan* [in Russian]. No 212-III of 9 January 2007, with amendments as of 26 December 2019. Available via https://online.zakon.kz/document/?doc_id=30085593. Accessed 9 January 2020.

Endo, A., Tsurita, I., Burnett, K., & Orencio, P. M. (2017). A review of the current state of research on the water, energy, and food nexus. *Journal of Hydrology: Regional Studies, 11,* 20–30.

Energy Information Agency (EIA). (2017). *Country analysis brief: Kazakhstan.* Available via https://www.eia.gov/international/content/analysis/countries_long/Kazakhstan/archive/pdf/Kazakhstan_2017.pdf. Accessed 2 May 2019.

Food and Agriculture Organisation of the United Nations (FAO). (2017). *Country fact sheet on food and agriculture policy trends. Socio-economic context and role of agriculture: Kazakhstan*. Available via http://www.fao.org/3/a-i7676e.pdf. Accessed 28 Mar 2020.

Food and Agriculture Organisation of the United Nations (FAO). (2018). *Water Accounting for Water Governance and Sustainable Development* (White Paper). Available via http://www.fao.org/3/I8890EN/I8890EN.pdf. Accessed 28 Mar 2020.

Food and Agriculture Organisation of the United Nations (FAO). (2012). *AQUASTAT-FAO's Information System on Water and Agriculture. Country profile: Kazakhstan*. Available via http://www.fao.org/aquastat/en/countries-and-basins/country-profiles/country/KAZ. Accessed 28 Mar 2020.

Government Decree. (2005). *Programme on Combating Desertification in the Republic of Kazakhstan for 2005–2015* [in Russian]. Decree No 49 of 24 January 2005. Available via https://tengrinews.kz/zakon/pravitelstvo_respubliki_kazahstan_premer_ministr_rk/ohrana_i_ispolzovanie_zemel/id-P050000049_/. Accessed 10 April 2020.

Government Decree. (2009). *On the draft of the National Plan for Integrated Water Resources Management and Increasing Water Use Efficiency for the period 2009–2025* [in Russian]. Decree of the Government of the Republic of Kazakhstan No 67 of 28 January 2009. Available via https://online.zakon.kz/Document/?doc_id=30379911. Accessed 4 April 2020.

Government Decree. (2010). *On the Water Programme "Ak Bulak" for the period 2011–2020* [in Russian]. Decree of the Government of the Republic of Kazakhstan No 1176 of 9 November 2010 Available via https://online.zakon.kz/document/?doc_id=30849574. Accessed 13 April 2020.

Government Decree. (2018). *State Programme for the Development of Agriculture in Kazakhstan for 2017–2021* [in Russian]. Decree No 423 of 12 July 2018. Available via https://online.zakon.kz/document/?doc_id=32033682. Accessed 10 April 2020.

Hoff, H. (2011). *Understanding the nexus*. Background paper for the Bonn 2011 conference: The water, energy and food security nexus. Stockholm Environment Institute, Stockholm. Available via https://www.scribd.com/document/318464767/Understanding-the-Nexus. Accessed 5 May 2020.

Howarth, C., & Monasterolo, I. (2016). Understanding barriers to decision making in the UK energy-food-water nexus: The added value of interdisciplinary approaches. *Environmental Science & Policy, 61*, 53–60.

Intergovernmental Panel on Climate Change (IPCC). (2007). *Climate change 2007: Synthesis report. Contribution of working groups I, II, and III to the fourth assessment report of the intergovernmental panel on climate change*. Cambridge: Cambridge University Press.

International Energy Agency (IEA). (2012). *World energy outlook*. Paris: OECD Publishing. Available via https://doi.org/10.1787/weo-2012-en. Accessed 28 March 2020.

Janusz-Pawletta, B. (2015). Current legal challenges to institutional governance of transboundary water resources in Central Asia and joint management arrangements. *Environmental Earth Sciences, 73*(2), 887–896.

Karatayev, M., Hall, S., Kalyuzhnova, Y., & Clarke, M. (2016). Renewable energy technology uptake in Kazakhstan: Policy drivers and barriers in a transitional economy. *Renewable and Sustainable Energy Reviews, 66,* 120–136.

Kazakhstani Electrical Grid Operating Company (KEGOC). (2018). *Annual report, 2018*. Available via https://www.kegoc.kz/en/shareholders-and-investors/information-disclosure/annual-reports/2018. Accessed 28 March 2020.

Kazakhstani Institute of Geography (KIG). (2013). *National atlas. Environment and natural resources*. Available via https://ingeo.kz/?p=2385. Accessed 28 March 2020.

KAZENERGY (Kazakhstan Association of Oil-Gas and Energy Sector Organisations). (2017). *The national energy report*. Available via http://www.kazenergy.com/upload/document/energy-report/National_Energy_Report-ENGLISH_03.09.pdf. Accessed 4 March 2019.

Kerimray, A., Kolyagin, I., & Suleimenov, B. (2018). Analysis of the energy intensity of Kazakhstan: from data compilation to decomposition analysis. *Energy Efficiency, 11*(2), 315–335.

Kerimray, A., Rojas-Solórzano, L., Torkmahalleh, M. A., Hopke, P. K., & Gallachóir, B. P. Ó. (2017). Coal use for residential heating: Patterns, health implications and lessons learned. *Energy for Sustainable Development, 40,* 19–30.

Khan, S., & Hanjra, M. A. (2009). Footprints of water and energy inputs in food production—Global perspectives. *Food Policy, 34*(2), 130–140.

Koshim, A., Karatayev, M., Clarke, M. L., & Nock, W. (2018). Spatial assessment of the distribution and potential of bioenergy resources in Kazakhstan. *Advances in Geosciences, 45,* 217–225.

Koulouri, A., & Mouraviev, N. (2019a). Introduction: The water-energy-food nexus through the lens of relational equity management. In A. Koulouri & N. Mouraviev (Eds.), *Policy and governance in the water-energy-food nexus: A relational equity approach* (pp. 1–7). Abingdon: Routledge.

Koulouri, A., & Mouraviev, N. (2019b). Governance of the water-energy-food nexus: A relational equity approach. In A. Koulouri & N. Mouraviev (Eds.), *Policy and governance in the water-energy-food nexus: A relational equity approach* (pp. 11–30). Abingdon: Routledge.

Land Code. (2003). *Land Code of the Republic of Kazakhstan* [in Russian]. Law No 442-II of 20 June 2003, with amendments as of 10 January 2020. Available via https://online.zakon.kz/document/?doc_id=1040583. Accessed 10 April 2020.

Law of the Republic of Kazakhstan. (2004). *On the Electrical Power Industry* [in Russian]. Law No 588-II of 9 July 2004. Available via https://online.zakon.kz/document/?doc_id=1049314. Accessed 9 January 2020.

Law of the Republic of Kazakhstan. (2009). *On the Utilisation of Renewable Energy Sources* [in Russian]. Law No 165-IV of 4 July 2009, with amendments as of 28 December 2018. Available via https://online.zakon.kz/Document/?doc_id=30445263. Accessed 7 April 2020.

Law of the Republic of Kazakhstan. (2012). *On Energy Saving and Increasing Efficiency of Energy Utilisation* [in Russian]. Law No 541-IV of 13 January 2012, with amendments as of 15 January 2019. Available via https://online.zakon.kz/Document/?doc_id=31112351. Accessed 9 January 2020.

Law of the Republic of Kazakhstan. (2016a). *On the Ratification of the Paris Agreement* [in Russian]. Law No 20-VI 3PK of 4 November 2016. Available via https://zakon.uchet.kz/rus/docs/Z1600000020. Accessed 2 April 2020.

Law of the Republic of Kazakhstan. (2016b). *On the Utilisation of Nuclear Energy* [in Russian]. Law No 442-V ЗРК of 12 January 2016. Available via https://zakon.uchet.kz/rus/docs/Z1600000442. Accessed 2 April 2020.

Leck, H., Conway, D., Bradshaw, M., & Rees, J. (2015). Tracing the water-energy-food nexus: Description, theory and practice. *Geography Compass, 9*(8), 445–460.

Long, A., Deng, M. J., Li, X., Zhang, Y., & Lei, Y. (2010). An analysis of water resources development in Kazakhstan and water problems. *Advances in Earth Science, 25*(12), 1357–1366.

Meng, E., Longmire, J., & Moldashev, A. (2000). Kazakhstan's wheat system: Priorities, constraints, and future prospects. *Food Policy, 25*(6), 701–717.

O'Hara, S., Hannan, T., & Genina, M. (2008). Assessing access to safe water and monitoring progress on MDG7 target 10 (access to safe water and basic sanitation): Lessons from Kazakhstan. *Water Policy, 10*(1), 1–24.

Organisation for Economic Co-operation and Development (OECD). (2019). *Monitoring the development of agricultural co-operatives in Kazakhstan*. Available via http://www.oecd.org/eurasia/competitiveness-programme/central-asia/Kazakhstan-Monitoring-Agricultural-Co-operatives-2019-EN.pdf. Accessed 28 March 2020.

Presidential Decree. (1999). *On the signing of the Kyoto Protocol by the Republic of Kazakhstan in relation to the United Nations framework convention on climate change* [in Russian]. Decree of the President of the Republic of Kazakhstan No 84 of 12 March 1999. Available via https://zakon.uchet.kz/rus/docs/U990000084_. Accessed 9 January 2020.

Presidential Decree. (2013). *On the concept of Kazakhstan's transition to a "green economy"* [in Russian]. Decree of the President of the Republic of Kazakhstan No 577 of 30 May 2013, with amendments of 10 September 2019. Available via https://online.zakon.kz/Document/?doc_id=31399596. Accessed 9 January 2020.

Rasul, G. (2016). Managing the food, water, and energy nexus for achieving the sustainable development goals in South Asia. *Environmental Development, 18,* 14–25.

Rasul, G., & Sharma, B. (2016). The nexus approach to water-energy-food security: An option for adaptation to climate change. *Climate Policy, 16*(6), 682–702.

Rhoe, V., Babu, S., & Reidhead, W. (2008). An analysis of food security and poverty in Central Asia-case study from Kazakhstan. *Journal of International Development: The Journal of the Development Studies Association, 20*(4), 452–465.

Rivotti, P., Karatayev, M., Mourão, Z. S., Shah, N., Clarke, M. L., & Konadu, D. D. (2019). Impact of future energy policy on water resources in Kazakhstan. *Energy Strategy Reviews, 24,* 261–267.

Salnikov, V., Turulina, G., Polyakova, S., Petrova, Y., & Skakova, A. (2015). Climate change in Kazakhstan during the past 70 years. *Quaternary International, 358,* 77–82.

Sinnott, P. (1992). The physical geography of Soviet Central Asia and the Aral Sea problem. In R. Lewis (Ed.), *Geographic perspectives on Soviet Central Asia* (pp. 73–98). London: Routledge.

Tankibayeva, A., & Adibayeva, A. (2019). Relational equity approach in the implementation of the Aral Sea basin Restoration Programmes. In A. Koulouri & N. Mouraviev (Eds.), *Policy and governance in the water-energy-food nexus: A relational equity approach* (pp. 95–119). Abingdon: Routledge.

United Nations (UN). (2011). *World population prospects: The 2010 revision. Volume I: Comprehensive tables.* New York: United Nations. Available via https://www.un.org/en/development/desa/population/publications/pdf/trends/WPP2010/WPP2010_Volume-I_Comprehensive-Tables.pdf. Accessed 28 March 2020.

United Nations (UN). (2014). *World urbanisation prospects: The 2014 revision.* Highlights. New York: United Nations. Available via https://population.un.org/wup/Publications/Files/WUP2014-Highlights.pdf. Accessed on 28 March 2020.

United Nations Development Programme (UNDP). (2004). *Water resources of Kazakhstan in the new millennium.* New York: United Nations Development Programme. Available via http://waterwiki.net/images/a/ad/KazakhstanWater.pdf. Accessed 30 March 2019.

Wang, X., Zheng, H., Wang, Z., Shan, Y., Meng, J., Liang, X., et al. (2019). Kazakhstan's CO_2 emissions in the post-Kyoto Protocol era: Production- and consumption-based analysis. *Journal of Environmental Management, 249,* 109–393.

Water Code of the Republic of Kazakhstan. (2003). *No 481-II of 9 July, with amendments as of 28 October 2019* [in Russian]. Available via https://online.zakon.kz/document/?doc_id=1042116. Accessed 12 April 2020.

World Bank. (2020). *Kazakhstan—Europe and Central Asia—P130013—Kazakhstan energy efficiency project—Procurement plan.* Available via http://documents.worldbank.org/curated/en/938821580197975175/pdf/Kazakhstan-EUROPE-AND-CENTRAL-ASIA-P130013-Kazakhstan-Energy-Efficiency-Project-Procurement-Plan.pdf. Accessed 28 March 2020.

World Economic Forum. (2011). *Global risks 2011* (Report). Available via http://www3.weforum.org/docs/WEF_Global_Risks_Report_2011.pdf. Accessed 29 April 2020.

World Resources Institute (WRI). (2016). *Water-energy nexus: Business risks and rewards* (Report). Available via https://wriorg.s3.amazonaws.com/s3fs-public/Water-Energy_Nexus_Business_Risks_and_Rewards.pdf. Accessed 28 March 2020.

World Resources Institute (WRI). (2019). *Ranking the world's most water-stressed countries in 2040.* Available via https://www.wri.org/blog/2015/08/ranking-world-s-most-water-stressed-countries-2040. Accessed 28 March 2020.

Zona KZ. (2020). *The concept of water resources management for 2020–2030 is approved in Kazakhstan 28 January* [in Russian]. Zona KZ (Internet newspaper). Available via https://zonakz.net/2020/01/28/koncepciya-programmy-upravleniya-vodnymi-resursami-na-2020-2030-gody-utverzhdena-v-rk/. Accessed 11 February 2020.

CHAPTER 8

Monotowns of Kazakhstan: Development Challenges and Opportunities

Madina Junussova and Aigul Beimisheva

Introduction

The *Kazakhstan-2050* strategy proposed by the nation's First President states that by 2050 the country aims to become one of the 30 most developed countries and increase the share of its urban population from 58% to 70% (Nazarbayev 2012). The government intends to use selected cities including Aktobe, Almaty, Nur-Sultan (formerly Astana), and Shymkent as drivers of the national economy (Government Resolution of the Republic of Kazakhstan No 728 2014). However, 15% of the country's urban population continues to live in industrial towns defined by the government as *monotowns* owing to their dependence on a single industry, primarily mining. Kazakhstan's economic dependence

M. Junussova (✉)
Institute of Public Policy and Administration,
University of Central Asia, Tekeli, Kazakhstan

A. Beimisheva
Department of Sociology and Social Work, Narxoz University,
Almaty, Kazakhstan

A. Koulouri and N. Mouraviev (eds.),
Kazakhstan's Developmental Journey,
https://doi.org/10.1007/978-981-15-6899-2_8

on natural resources makes these towns important for the national development. In 2019, the mining industry contributed 16% of the country's gross domestic product (GDP) (Statistics Department of the Eurasian Economic Committee 2020). In the pursuit of urbanisation, the government is expected to make efforts to improve economic conditions not only in large cities, but also in monotowns. Therefore, Kazakhstan's urbanisation policy faces the question of how to balance urban and regional development to benefit not only large cities, but also smaller towns and villages.

Since 2012, Kazakhstan's government has adopted a range of policies and made considerable budget transfers for the monotowns' development. The national government assisted monotowns by subsidising the operations of the main industries and upgrading roads and municipal infrastructure, but the national support has thus far failed to supply these towns with a strong economic basis that will ensure their long-term sustainability. The purpose of this chapter is to provide a critical review of Kazakhstan's monotown policy and propose the changes needed to assist monotowns with finding their own sustainable economic development paths. The chapter begins with an overview of the transition of mining towns from dealing with the Soviet legacy to facing market realities and their current development challenges. Then the chapter provides an assessment of national government actions aimed to support monotowns. The analysis of the impact of monotown policies is based on the findings of field interviews, reviews of monotown development plans, and local statistics. The details of the study participants are shown in Table 8.5 (see Appendix). The policy impact is illustrated using examples of different monotowns in Kazakhstan by contrasting achievements with missed opportunities. Based on the findings, the authors suggest the changes that can help monotowns thrive.

The Transition of Monotowns from the Soviet Legacy to a Market Economy

Kazakhstan's mining towns are the results of the Soviet industrialisation policy that included the intensive development of chemical, coal, and petroleum industries and metallurgy, and the establishment of towns to serve these industries. The location of towns was largely determined by the presence of valuable natural mineral resources. Some towns were set up to serve the extraction of metal ores; these towns include Arkalyk,

Balkhash, Karazhal, Kentau, Khromtau, Lisakovsk, Ridder, Rudny, Tekeli, and Zyryanovsk. Rudny, in the Kostanay region, was built in 1957 to facilitate mining of the largest iron ore deposit in the country (Akimat of Rudny 2019). Tekeli, in the Almaty region, was founded at the site of a large polymetallic deposit in the late 1930s. Most mining towns of the Soviet Kazakhstan experienced fast growth during World War II when industries were relocated from the north to the south of the Soviet Union. At that time towns such as Tekeli and Kentau ramped up their lead mining to ensure supply to the wartime weapons production industries (Zholdybayev and Zhamelov 2007). During the Soviet time, most of the mining towns were closed to the visitors and developed in isolation from the surrounding regions. Supervised directly by the ministries located in Moscow, monotowns benefited from the centrally organised direct supply of consumer goods and labour resources.

During the Soviet time, the populations of mining towns were supplemented with dwellers from other parts of the country as a result of the forced resettlement policy. For example, Kentau, in the Turkestan region, was called *little Athens* due to the large number of Greeks living there (Trotchenko 2018). In the Soviet era, some Kazakhstani monotowns became home to many Koreans (Kim 1993). Many ethnic Germans, Greeks, Koreans, and Russians were deported to work in Karatau, where some districts were also called *little Athens* and *West Berlin* (Participant 44). Japanese prisoners of World War II were sent to Tekeli to work on the construction of industrial assets and residential areas (Akhmediyeva 2012). During the post-war restoration, many ethnic German and Poles were sent to work in the mining towns of Soviet Kazakhstan (Vechor-Shcherbovich 1975). Educated and qualified specialists preferred to move from the depressing mining towns of Siberia to mining towns located in Kazakhstan and other southern regions with a better climate and living conditions: 'In the 1960s, our town was famous for having well-paid jobs and offering comfortable living. Many Soviet specialists in medicine, engineering and education moved to Tekeli and our parents were among them' (Participant 32).

Soviet mining towns were often called *monotowns* because of their economic dependence on the operation of a single industry. In the Soviet time, the main industry acted not only as the main provider of jobs, but also of social infrastructure, municipal utilities, and housing for its employees (BISAM Central Asia 2012). The main industrial enterprise and its administration played a key role in town management and development as the industry was the main provider of almost all town services. A mining plant typically owned all urban infrastructure, from industrial facilities to restaurants and cultural centres, while the town's administration was the local executor of top-down orders from the Ministry of Defence and the Ministry of Geology of the Soviet Union.

The dependency of the monotowns on a single mining industry made them vulnerable to the industry's decline when the country transitioned from the planned Soviet system to a market-oriented economy. When Kazakhstan gained independence in 1991, most local industries lost their technological links to Russia and other former Soviet countries. Kazakhstan's mining industries lost their main consumers and had to find new markets to sell their products. In some towns, the collapse of the Soviet Union occurred simultaneously with the inevitable mine exhaustion. For example, in Tekeli, up to 45% of lead deposits were exhausted by the 1990s and the plant shut down completely in 1996 due to a shortage of polymetallic resources (Electromanganese 2020). Newly formed monotown authorities called *akimats* were weak and continued to be dependent on the dominant local industry and top-down orders from the regional and national government. 'In the Soviet past, the plant administration provided us with jobs and services. Now, we have a town authority that tries to solve town development issues, but it seems that the *Akimat* does not have sufficient budgetary resources or power to help us with employment and the provision of good quality public services' (Participant 48).

In the 1990s, the national government initiated the privatisation of industrial enterprises, in which monotown authorities could take part. Municipal utility networks formerly managed and owned by mining enterprises were partly privatised and partly transferred to the supervision of town authorities. In some cases, foreign companies became owners of local industrial enterprises that continued mining operations, but foreign firms were generally not interested in making considerable capital investments in the modernisation of industry or the monotowns' infrastructure. For example, in 1995, the Mittal Steel Company became the owner of the metallurgical plant in Temirtau (Akimat of Temirtau 2020). 'We live in a district that is still not connected to the central heating and water supply systems. The town authority promises to solve this problem, pointing to the unwillingness of our main industry to cooperate or invest in urban development. We have become victims of privatisation, and some districts in our town are better developed than others' (Participant 42). In another monotown—Khromtau, in the Aktobe region—the mining company owns and manages the municipal utility networks in most of the town districts, but not all residents have access to the water and heating infrastructure. 'I live in a new district and we do not have access to the centralised water and heating supply systems. Town authorities continue to promise us that the issue will be solved soon, but we continue to incur additional monthly expenses because we buy bottled water and use electric heaters' (Participant 43). To some extent, the privatisation of the municipal infrastructure of mining towns impacted negatively the maintenance of local utility networks and led to the fast degradation of housing. For example, 70–90% of the heat and water supply networks were worn out by 2012 in towns of the East-Kazakhstan region (BISAM Central Asia 2012). In Rudny, the deterioration level of the water and sewer networks reached 60% in 2018 (Akimat of Rudny 2016).

To prevent mass unemployment and social tension, Kazakhstan's government tried to support monotowns by providing special loans and subsidies to selected local enterprises. For example, in 1996, the Tekeli plant received a loan of KZT 200 million (USD 3 million) from the State Rehabilitation Bank of the Republic of Kazakhstan to replenish its operating funds and ensure uninterrupted payroll (Government Resolution of the Republic of Kazakhstan No 1187 199). By that time, the number of plant employees had decreased by 313 people. In 1997, the Tekeli plant received an additional loan of KZT 15 million (USD 203,000) to repay its debt to the Shakhtostroitel Corporation. The loans, including the one to the Tekeli plant, were extended from the national budget and aimed to rehabilitate insolvent enterprises. Nevertheless, mining companies could not operate at the same level of productivity or employ as many town residents as in the past. The government subsidies had a temporary effect and kept some enterprises afloat for only a few years.

The decline of local industries and the deterioration of the living conditions in the late 1990s led to an active resident outflow from the mining towns. 'I remember in the 1990s many people lost their jobs because of a plant closure, and former plant workers left town to find new jobs. Most of them moved to other regions of Kazakhstan and also to Russia where mining plants were operating or emigrated to foreign countries' (Participant 26). Most of the members of the populations who had been forcibly resettled to the mining towns of Soviet Kazakhstan returned to their homelands. The most dramatic out-migration occurred in Arkalyk,

Karazhal, Saran, and Shakhtinsk. In 2000, the population in Saran and Shakhtinsk had decreased by 35% compared to 1991, and in Arkalyk and Karazhak by 34% (Committee on Statistics under the Ministry of the National Economy of the Republic of Kazakhstan [Committee on Statistics of Kazakhstan] 2020). In 1999, 19% of the urban population lived in monotowns, which decreased to 14% by 2016 (Organisation for Economic Cooperation and Development [OECD] 2017). At the same time, some monotowns surrounded by rural areas experienced population growth due to the migration of residents of neighbouring villages who were attracted by the comparatively better developed urban infrastructure.

Development Problems of Monotowns

A demographic decline is one of the major development problems of monotowns as it leads to a shortage of qualified specialists. Out-migration of young people and skilled workers occurs in response to the decline of the mining industry, few alternative employment opportunities, and low wages. More than 10% of the population of monotowns are retirees (Committee on Statistics of Kazakhstan 2020). People between 18 and 30 years of age comprise less than 30% of the monotown population. The difference in salary levels is one of the important drivers of interregional migration in Kazakhstan (Aldashev and Dietz 2014). The average monthly salary in monotowns typically remains below the national level. In 2018, the average monthly salary in monotowns amounted to KZT 159,700 (USD 490), which is slightly below the national average of KZT 162,300 (USD 506). The average salary in 2018 in Saran and Shakhtinsk was below USD 300. The higher average monthly salary in some monotowns is associated with the oil and gas industries. For example, in 2018, the average monthly salary in Aksai was KZT 448,800 (USD 1402) and in Zhanaozen KZT 435,400 (USD 1361), which was even higher than the average monthly salary in the large cities such as Almaty and Nur-Sultan. Salaries account for 50–60% of household incomes in monotowns with the rest generally coming from self-employment (Participants 32–35).

Self-employment remains a key source of income for residents of monotowns. In 40% of monotowns, the self-employment rate is higher than in all other urban areas of Kazakhstan (Committee on Statistics of Kazakhstan 2020). The largest share of self-employed workers can be found in the post-mining monotowns, such as Kentau and Tekeli, where industrial enterprises have ceased mining operations, or where industrial enterprises are not in full operation, as in Arkalyk, Karatau, and Zhanatas. The smallest numbers of self-employed workers are in monotowns where the mining industry is in operation, such as in Kurchatov, Ridder, and Satpaev. Monotown residents generally consider self-employment a temporary activity while they seek a well-paid job (Beisembaev et al. 2017). For young people, unfavourable economic conditions and low levels of education are the main causes of self-employment. Self-employment, in many cases is related to working in the informal economic sectors that restricts people to benefit from labour regulations or social security schemes.

Compared to other urban areas, monotowns generally have a relatively small number of residents with higher levels of education. 'When I moved to Tekeli many people were surprised that I had a university degree. Later I learned that there are few people with higher education in the town simply because in the past most residents worked at the main plant and did not need a university degree' (Participant 26). Today, better educated workers tend to move from monotowns to larger cities. Most universities in Kazakhstan are based in large cities. University students often do not return to monotowns after graduation due to the lack of well-paid jobs. 'I graduated from a university in Almaty and came back here, but most of my friends who went away to study did not return. Of course, they were attracted by job opportunities and better living conditions available in large cities' (Participant 27). The lack of human capital hinders monotowns trying to attract private investors and develop new economic activities. 'We want to develop tourism, but we do not have skilled tourist guides who can speak English or other foreign languages. Many entrepreneurs from Almaty and Nur-Sultan are attracted by our beautiful mountains and natural surroundings, and want to build private resorts, but the main obstacle is the absence of qualified workers and managers' (Participant 4).

Since the 2000s, the government has been trying to fill some of the human capital gaps in monotowns by settling ethnic repatriates in mining towns. In 1992, former President Nazarbayev announced quotas for receiving ethnic repatriates, which increased from 500 families in 2000 to 20,000 in 2009 (Central Asia Monitor 2012). In many cases, the population growth in monotowns after 2010 was associated with the inflow of repatriates (Participants 27–30, 34–36). However, many new residents experienced difficulties finding adequate housing. In the 1990s, many people who were forcibly relocated to monotowns during the Soviet time returned to their homelands, often simply abandoning their monotown dwellings. New monotown residents could not move into the abandoned homes because it was necessary to get permission from the previous owners (BISAM Central Asia 2012). Many repatriates could not get local registration (*propiska*) because they were unable to secure their own officially registered housing. As a result, repatriates did not receive adequate access to most local public services or employment.

The Impact of Monotown Policies

For many years, the development of monotowns was primarily associated with a single enterprise around which a town was established and the government supported these towns by investing in modernising the industry. In 2003, the government adopted the *Strategy of Industrial and Innovative Development 2015*, aimed at increasing the share of value-added products in the country's export portfolio by developing manufacturing (Government Resolution of the Republic of Kazakhstan No 1096 2003). Some industries began technological updates, but modernisation led to considerable decrease in the workforce and salaries. 'We changed to a technological process that did not require as much labour, so of course we had to decrease the number of personnel. We mainly kept workers with engineering education who could learn how to work with new equipment without any additional investment in their training' (Participant 11). The situation worsened in 2009, after the global economic crisis and decreases in the world market price of oil and gas. The government reacted by the enactment, in 2010, of the national programme of *Fast Industrial and Innovative Development 2010–2014* and initiated the development of the *Industrialisation Map* (Government Resolution of the Republic of Kazakhstan No 958 2010). In total, monotown authorities proposed 114 projects to be included in the Industrialisation Map, at a cost of KZT 1249 billion (USD 8327 million), but by 2012, the government had approved only 81 projects at a cost of KZT 22 billion (USD 143 million) (Ministry of the Economic Development and Trade 2012). Unfortunately, the government subsidies did not reach all monotowns, whereas worsening living conditions led to increasing social tension. In 2011, industrial workers in Zhanaozen, not satisfied with their salaries and living conditions, went on strike that evolved into a violent conflict with the town authorities and police, leading to the death of 16 people (Kurmanov 2016).

From Industrial Policy to Monotown Policy: Delegation of Planning to Town Authorities

In 2012, a year after the Zhanaozen incident, the national government approved the first monotown development programme. To avoid conflict in other monotowns, the government conducted several studies and adopted the national programme for the *Development of Monotowns* (hereafter, Monotown Programme) (Government Resolution of the Republic of Kazakhstan No 683 2012). The Monotown Programme identified 27 monotowns as towns with at least one enterprise belonging to the same industry, and being responsible for the employment of at least 20% of the population and for the overall industrial output. Table 8.1 shows the full list of identified monotowns. There are eight of the 27 monotowns in the Karagandy region. There are four monotowns in the Kostanay and East Kazakhstan regions; all other regions have only one or two. The Monotown Programme aimed to assist towns with tackling economic development problems by (i) increasing the efficiency of the main operating industry, (ii) supporting economic diversification and small and medium-sized enterprises (SMEs), (iii) stimulating labour mobility, and (iv) developing social and engineering infrastructure.

Being officially defined as a monotown brought the selected towns opportunities to attract more public investment from the national budget. However, public investment was allocated according to a monotown's categorisation by economic potential. The Ministry of the National Economy proposed a unified methodology for assessing the economic potential of monotowns. Based on the ministry's methodology, monotowns are categorised as having high, medium, or low economic potential, and policy packages had to be tailored to each town's estimated economic potential (see Table 8.1). Monotowns were assigned to a category based on several criteria. Information on the weighting of each criterion in the overall evaluation methodology was not provided in the Monotown Programme documents or elsewhere. Nonetheless, the example of how the ministry assessed monotowns is provided in Table 8.2. Aksu, Ekibastuz, Ridder, Rudny, and Temirtau were identified as having high economic potential. Only Arkalyk and Zhanatas were categorised as having low economic potential. However, monotown residents argue that the real number of monotowns with low economic potential would have

Table 8.1 Monotowns' economic potential

No	*Name of town*	*Population in 2011*	*Economic potential*
Karagandy region			
1	Balkhash	77,056	Medium
2	Zhezkazgan	85,185	Medium
3	Karazhal	19,235	Medium
4	Saran	50,821	Medium
5	Satpaev	69,744	Medium
6	Temirtau	179,853	High
7	Shakhtinsk	56,263	Medium
8	Abay	26,249	Medium
Kostanay region			
9	Arkalyk	27,634	Low
10	Zhetykara	34,595	Medium
11	Lisakovsk	40,546	Medium
12	Rudny	126,613	High
East Kazakhstan region			
13	Zyryanovsk	38,185	Medium
14	Kurchatov	11,198	Medium
15	Ridder	54,793	High
16	Serebryansk	9653	Medium
Pavlodar region			
17	Aksu	45,845	High
18	Ekibastuz	136,829	High
Zhambyl region			
19	Karatau	27,317	Medium
20	Zhanatas	21,098	Low
Mangystau region			
21	Zhanaozen	121,327	Medium
South Kazakhstan region			
22	Kentau	61,362	Medium
Akmola region			
23	Stepnogorsk	64,316	Medium
Aktobe region			
24	Khromtau	24,617	High
Almaty region			
25	Tekeli	28,656	Medium
Atyrau region			
26	Kulsary	53,513	Medium
West Kazakhstan region			
27	Aksai	33,610	Medium

Source Compiled by the authors based on the Government Resolution of the Republic of Kazakhstan No 683 (2012)

Table 8.2 Criteria for the assessment of the monotowns' economic potential

	Criteria	*High potential*	*Low potential*	*Assessment of Arkalyk and Zhanatas*
1	The presence of a stable extractive or manufacturing industry and the availability of a raw material base for future mining	+	–	During the last ten years, the main enterprise reduced production and was not profitable The mineral base is expected to be exhausted within the next ten years
2	Budget self-sufficiency	Budget revenue is sufficient to cover expenses	Recipient of national and regional government budget transfers	Both towns are recipients of national and regional government budget transfers Average wages in the towns are lower than the regional average
3	The location of a town at the intersection of large transport corridors	+	–	Poor transport connectivity and road quality and remote location with more than an hour's drive or 75 km distance to/from main cities and transport corridors
4	Location of a town near a large city or being part of an agglomeration	+	–	
5	Potential for economic diversification	+	–	–
6	Possibility to provide a wide range of services to surrounding communities	+	–	–
7	High proportion of active SMEs	+	–	–

(continued)

Table 8.2 (continued)

	Criteria	*High potential*	*Low potential*	*Assessment of Arkalyk and Zhanatas*
8	Negative dynamics of the migration balance over the past ten years	–	+	The net migration rate has been negative, and emigration is high among skilled workers
9	Deterioration of engineering infrastructure	Acceptable level	Below standards	Worn-out municipal utility and electricity distribution networks, and outdated housing
10	Ecological situation	Acceptable level	Below standards	Environmental problems exist, air and water pollution levels are higher than acceptable levels, according to national environmental norms

Source Compiled by the authors based on the data of the Ministry of the National Economy (2012) been higher if the highest weightings were assigned to the criteria 'negative dynamics of the migration' and 'ecological situation' (Participants 18–21, 28–30, 35–38).

The Monotown Programme supplied town authorities with a local planning document. Each of the 27 monotown authorities had to develop *Comprehensive Development Plans* with estimations of the public investment required to upgrade local urban infrastructure (roads, municipal utility networks, and housing) and information about local investment projects. Comprehensive Development Plans served as a new basis for planning government financial support to monotowns from national and regional budgets. However, due to top-down administration, fiscal constraints, and the lack of planning autonomy and capacity, town administrators were unable to use the new planning tool successfully. Despite the efforts of the Ministry of the National Economy to promote the design of Comprehensive Development Plans in towns starting in 2012, by 2015 seven out of the 27 town authorities (Arkalyk, Khromtau, Kulsary, Rudny, Saran, Serebryansk, and Zyryanovsk) had still not approved their plans, explaining that they remained 'work in progress' (Participant 1).

Monotowns in Regional Development Policy: Success and Failure of Industry-Biased Investment

In 2014, the Monotown Programme became an integral part of the *Regional Development Programme 2020* now governing the development of all Kazakhstan's urban areas (Government Resolution of the Republic of Kazakhstan No 728 2014). In fact, the government merged the Monotown Programme with other ongoing programmes including the *Modernisation of Housing and Utilities*, *Akbulak 2020* (aimed at improving the supply of drinking water) and *Affordable Housing 2020*, without ensuring adequate coherence between all policy actions (OECD 2017). Furthermore, the Regional Development Programme 2020 introduced certain conflicting centralisation and decentralisation efforts. Although some efforts supported the development of large urban agglomerations by attracting human capital from monotowns, investment in monotowns remained focused on limiting the outflow of their population.

Town authorities had to propose investment projects supported by the main industries operating in monotowns (also called *anchor* projects);

these projects were included in the national Industrialisation Map. The Ministry of the National Economy wanted monotown authorities to attract investment from national and foreign companies. Some monotowns with large operational enterprises were able to engage industry in the town's development (BISAM Central Asia 2012). In Arkalyk, in 2015, the main industrial enterprise Aluminstroy LLC invested KZT 10.7 million (USD 59,000) in the repair of a section of the local road from Arkalyk to the village of Shili, the preparation of foundations for gravel roads, and the mowing of grass on the roadsides (Akimat of Arkalyk 2016). In Zhanatas, the town authorities benefitted from a foreign investor who contributed to the renewal of the local industry (Akimat of Zhanatas 2016). The Russian company EuroChem launched a chemical complex with the capacity to produce 1 million tons of fertilisers using deposits from the phosphorite basin of Karatau. The total investment of EuroChem was estimated at USD 750 million. In 2016, EuroChem engaged in urban regeneration by ordering a town development master plan (Suilemanova et al. 2017). The financial support of EuroChem allowed the town to demolish old multi-storey apartment buildings and renovate most of its degrading housing.

Many other monotowns failed to attract public and private investment. In 2015, fewer than the half of the 48 projects proposed by monotowns in 2012 had been implemented 'due to the technical weakness of the projects and the lack of coordination between local entrepreneurs and town authorities' (Participant 3). By 2019, only 32 of the 48 projects had been implemented and new start-ups were launched making use of received funding; seven projects were suspended; and nine enterprises began working at lower capacity due to a shortage of human resources (Government Resolution of the Republic of Kazakhstan No 990 2019). Authorities in Kurchatov, Ridder, and Zyryanovsk failed to attract foreign investment because of the lack of qualified staff (Participant 1): it was difficult to find engineers to launch high-tech companies.

Initially, investment projects proposed by monotowns were required to assist economic restructuring by developing alternative activities to replace the mining industry and combat growing unemployment. However, as noted by the national government, none of the projects was innovative or added to the diversification of the local economy (Government Resolution of the Republic of Kazakhstan No 990 2019). Instead, the proposed projects aimed at the extraction or primary processing of raw materials, such as lime, sand, and gravel, and the production of paving and building materials. Furthermore, most of the investment projects did not make any considerable impact on decreasing unemployment. For example, in Stepnogorsk between 2010 and 2018 KZT 3.5 billion (USD 20 million) were invested in the reconstruction of a sulphuric acid plant, the production of glyphosate potassium salt and glyphosate- containing herbicides, and the reconstruction of mini-refineries using catalytic processing of petroleum products (Akimat of Stepnogorsk 2017). The implementation of these industry-related investment projects created only 351 jobs, which did not significantly decrease unemployment numbers. In 2014–2016, the number of unemployed workers in the town still was about 1700.

Local Implementation Challenges: Lack of Public–Private Dialogue

The Regional Development Programme 2020 replaced the Comprehensive Development Plans with the *Territorial Development Programmes*. Monotown authorities again struggled to produce feasible Territorial Development Programmes and support innovative investment projects because 'town authorities lack expert knowledge and they could not even follow the Ministry of the National Economy instructions' (Participant 22). The unified planning approach did not provide enough flexibility to adapt suggested policy measures to local circumstances (OECD 2017). As in the case of the Russian monotowns, a top-down universal approach to determining quantitative indicators for all towns did not allow local administrators to make use of the unique competitive advantages of their locality (Didyk and Ryabova 2014).

The ministry did not provide adequate resources, but strongly suggested that all monotowns followed a single methodology with an approved list of target indicators. Similarly to the other 26 town authorities of the country, the Balkhash town authorities included in their Territorial Development Programme 2016–2020 almost all development objectives and target indicators recommended by the ministry (see Table 8.3). Ignoring the unique location of the town at the shore of the Balkhash Lake, the town authorities did not propose any location-specific development objective that could be useful for mobilising the town's unique advantages. Moreover, there was a shortage of qualitative indicators assessing the quality of life from the perspective of the Balkhash town residents. The selected social indicators did not provide any evidence about the satisfaction of the Balkhash residents with the quality of locally

Table 8.3 Territorial Development Programme of the town of Balkhash: Objectives and indicators set by the Ministry of the National Economy

Development objectives	*Development indicators*
Economic development • To diversify and improve the competitiveness of the manufacturing industry • To enhance the investment climate • To increase the competitiveness of agricultural products • To create favourable conditions for the development of small and medium-sized businesses	• The volume of manufacturing output • The volume of output of the metallurgical industry • The volume of output of the mining and quarrying industries • The amount of investment in fixed assets • The amount of investment in manufacturing • The share of innovative enterprises • The volume of gross agricultural production • The amount of investment in fixed assets in agriculture • The volume of output of small and medium-sized enterprises • The amount of manufactured products and services provided by medium-sized enterprises

Development objectives	*Development indicators*
Social development • To ensure a high quality and good accessibility of education • To increase the effectiveness of the protection of children's rights • To improve public health and ensure sustainable demographic development • To create effective systems for the social protection of citizens • To promote the culture, sport, and tourism industries • To strengthen the use of the state language and develop multilingualism	• The number of functioning emergency and three-shift schools • The enrolment of children (ages 3–6 years) into preschool education • The rates of maternal and infant mortality • The availability of primary health care within walking distance • The unemployment rates and youth unemployment • The percentage of people with disabilities out of the total number of employees • The number of jobs created • The share of social and transport infrastructure facilities with access for people with disabilities • The average number of visitors (visits) to cultural centres, libraries, and leisure facilities • The availability of sports infrastructure • The growth of the number of visitors • The share of the adult population that speaks Kazakh, Russian, and English

(continued)

Table 8.3 (continued)

Development objectives	*Development indicators*
Infrastructure development • To ensure the sustainable growth of the construction sector • To provide road networks • To improve the quality of communication services • To enhance the availability of utilities infrastructure and housing	• The volume of construction work • The share of local roads with surfaces in good or satisfactory condition • The number of Internet users • The level of digital literacy • The provision of housing per resident • The share of communal facilities requiring major repairs • The availability of centralised water supply
Environmental and land use conditions • To ensure environmental safety and improve urban hygiene and sanitary conditions	• The volume of processed municipal solid waste • The share of facilities for the disposal of municipal solid waste that meet environmental requirements and sanitary rules • The availability of wastewater treatment

Source Adapted by the authors from the Territorial Development Programme of the Balkhash town (Akimat of Balkhash 2016) provided public services. None of the selected economic and infrastructure development indicators were useful in assessing local conditions for doing business in Balkhash. Despite the critical ecological situation in the Balkhash town, the last objective aiming to improve environmental conditions was not supported by important indicators (i.e. those with higher weighting) aiming to measure the levels of air, water, and soil pollution.

Overall, the Territorial Development Programmes serve as directive administrative tools rather than workable local strategies. Monotowns were unable to use the Territorial Development Programme format to find solutions to local environmental problems negatively impacting any promising potential for the development of tourism. For example, the Kentau and Tekeli administrators complained that they could not do anything with former industrial lands because they still belonged to the respective ministries that were not interested in the rehabilitation of brownfields (Participants 7 and 9). In addition, the town administrators complained about the lack of locally available regulatory power to foster local business development (Participants 8 and 10). For example, the Territorial Development Programmes helped the Tekeli authorities to create some jobs, provide grants to local businesses, and repair municipal infrastructure. However, they were unable to use the territorial programmes to improve the local business environment.

Compared to previous monotown programmes, the Regional Development Programme 2020 put greater emphasis on fostering entrepreneurship. In 2015, the national budget allocated KZT 22 billion (USD 7 million) to support entrepreneurship by providing a low-interest loan that was given in addition to regular budget transfers to monotowns (OECD 2017). Monotown authorities distributed grants ranging from KZT 1.5 million to 3 million (USD 6700–13,530) to promising new business projects. Monotown entrepreneurs were also given special training and individual consultations. However, the local monotown business people complained about their inability to qualify for the loans due to their lack of collateral. Also, loans and grants were too small for launching a sustainable business. Furthermore, there were complaints that people responsible for training and consultations were not sufficiently qualified to provide adequate skills training and competent advice (Participants 23–30).

The national government did not offer any proper instructions regarding how town administrators should engage local private and public actors and various interest groups in the planning of the towns' future.

However, the Ministry of the National Economy wanted town authorities to hold effective public discussions with private firms and government agencies to identify the best investment projects. Local entrepreneurs and industry representatives of many monotowns complained that 'nobody invited us to participate in the development of the Territorial Development Programmes' and 'local investment projects were not publicly discussed' (Participants 25 and 30). For example, in the East Kazakhstan region, 50% of interviewed entrepreneurs stated that they were invited to the discussion of investment projects or town development plans, whereas only 33% of entrepreneurs interviewed in Serebryansk, 13% in Kurchatov, and 10% in Zyryanovsk confirmed the same (BISAM Central Asia 2012).

The lack of participatory planning tools and administrative power did not allow town authorities to develop workable Territorial Development Programmes that incorporated market realities and the development priorities of local businesses. For example, the Ministry of the National Economy wanted town administrators to regulate local markets by asking local industrial enterprises to order goods and services from businesses located in the area. However, the national government did not provide administrators with workable mechanisms to help local businesses increase the quality of goods and services to meet the needs of the industry. Most of the large enterprises refused to channel orders to local companies, preferring to purchase work clothes, food, and other goods and services on the wider market, from not only national, but also international companies (BISAM Central Asia 2012). The industry representatives complained about the low quality of local goods and services (Participants 11–14). In turn, the local business representatives complained that town authorities could not enforce the request that large enterprises should buy local goods. 'I think the problem was in the absence of cooperation between the industry and the town authorities who were then unable to propose alternative solutions for how to support local business' (Participant 22). Constrained by the centralised public procurement system rules, town administrators could not even place their own public orders with local producers (Participants 8 and 10).

Path Dependence Symptoms: Subsidised Development Constrained by Centralised Decision-Making

The continuing adjustments to the Regional Development Programme 2020 prevented its proper implementation and created uncertainty about the future of monotowns. Since 2014, the Regional Development Programme 2020 was amended multiple times: four times in 2015 (Government Resolutions of the Republic of Kazakhstan No 15 2015; 42 2015; 284 2015; 451 2015), then in 2016 (Government Resolution of the Republic of Kazakhstan No 499 2016) and again in 2018 (Government Resolution of the Republic of Kazakhstan No 767 2018). Tracking the programme's progress using the set indicators, such as the number of SMEs, the volume of production from manufacturing industries, and the unemployment rate showed that not all monotowns were able to achieve the desired diversification for their economic development. The share of active businesses among the registered SMEs increased from 39% in 2012 to 79% in 2017 (Committee on Statistics of Kazakhstan 2020). However, SMEs continue to experience growth-related problems and contribute on average less than 1% of the monotowns' economic output (Committee on Statistics of Kazakhstan 2020). Compared to 2012, the volume of production from manufacturing industries in 2017 increased in most monotowns except in Zyryanovsk where it declined by 60.9% and Ekibastuz where it declined by 13.8% (Committee on Statistics of Kazakhstan 2020). The unemployment rate in monotowns decreased from 5.5% in 2014 to 5% in 2017, but the unemployment rate among young people (aged 15–24) increased from 4.5 to 7% (Committee on Statistics of Kazakhstan 2020).

Launched in 2019, the *Regional Development Programme 2025* introduced a new categorisation of monotowns, but it did not change the established subsidy-based approach that kept an exclusive focus on a single industry making the monotowns' economies even more dependent on the extraction of natural resources (Government Resolution of the Republic of Kazakhstan No 990 2019). The Ministry of the National Economy proposed to differentiate the government efforts supporting monotowns. The new categorisation identified towns that could benefit from the agglomeration effect because of their proximity to large cities (Abay, Aksu, Kentau, Khromtau, Ore, Saran, Tekeli, and Temirtau) and towns located far from large cities (Aksai, Altai, Arkalyk, Balkhash, Ekibastuz, Karatau, Karazhal, Kulsary, Kurchatov, Lisakovsk, Ridder, Satpaev, Serebryansk, Shakhtinsk, Stepnogorsk, Zhanaozen, Zhanatas, Zhezkazgan, and Zhitikara). However, regarding monotowns located close to urban agglomerations, the national government continues to push the restoration of the previous narrow specialisation of the main large enterprise. As for the remote monotowns, the ministry again plans to evaluate their development potential and suggest development priorities for five, ten, and twenty years, applying a top-down planning approach. Then, based on the results of the government's evaluation, monotown authorities will develop proposals for economic development, which will be included in the Territorial Development Programmes for seeking funding from the national budget.

The Regional Development Programme 2025 appeared not much different from the 2020 version. It is unlikely that it will help monotowns overcome the programme implementation challenges resulting from the existing institutional constraints, such as the centralised public administration and the lack of fiscal freedom. Monotowns belong to the lowest levels of the subnational government and they do not have decision-making powers over planning and management of their development. For example, seven out of the eight monotowns in the Karagandy region have the status of 'towns of regional importance' and belong to the second tier of the subnational government, subordinated to the Karagandy region Akimat, whereas Abay belongs to the third tier of the subnational government and is subordinate to both the Abay district and the Karagandy region. Table 8.4 shows the different levels of subnational government. In the case of the Zhambyl region, both Zhanatas and Karatau belong to the third tier of the subnational government and are subordinate to the Sarysu and Talas districts (see Table 8.4). According to the national planning system, only monotowns belonging to the second tier of the subnational government participate in local development planning by drafting Territorial Development Programmes that enable them to qualify for special allocations from the regional budget. With the replacement of the Monotown Programme by the Regional Development Programmes 2020 and 2025, Territorial Development Programmes had to be developed instead of Comprehensive Development Programmes, meaning that monotowns belonging to the third tier of subnational government ceased to produce any local development plans. The monotowns belonging to the lowest tier of subnational government do not have their own budgets or revenue sources (see Table 8.4).

In Kazakhstan, there are three levels of budget—national, regional, and district—meaning that some monotowns do not have a local budget. In 2015, Kazakhstan's President announced a plan that includes 100 steps towards implementing five institutional reforms. These reforms include forming a modern state apparatus, ensuring the rule of law, fostering industrialisation and economic growth, building a nation of a united

Table 8.4 Public administration, planning, and budget structure of Kazakhstan

Main tiers of public administration		*Key planning documents*	*Budget system*
National government Ministry of the National Economy		State Regional Development Programme 2020–2025	National budget
Regional government: First tier of the subnational government		Five-year Territorial Development Programme E.g. Karagandy region Territorial Development Programme 2020–2025	Three-year Regional budget
Executive body *Akimat* develops local plans and budgets	Representative body *Maslikhat* approves local plans and budgets		
District-level government: Second tier of the subnational government		Five-year Territorial Development Programme E.g. Balkhash, Zhezkazgan, Karazhal, Saran, Satpaev, Shakhtinsk, and Temirtau in the Karagandy region	Three-year District budget
Town *Akimat* subordinate to the regional *Akimat* and are responsible for developing local plans and budgets	Town *Maslikhat* subordinate to the regional *Maslikhat* and are responsible for approving local plans and budgets		

(continued)

Table 8.4 (continued)

Main tiers of public administration	*Key planning documents*	*Budget system*
Rural-level government: Third tier of the subnational government There is no special authority, only the head of the town or village called *Akim*	No planning documents E.g. Abay town of the Abay district of the Karagandy region E.g. Zhanatas town of the Sarysu district and Karatau town of the Talas district of the Zhambyl region	No local budget

Source Compiled by the authors

future, and strengthening a transparent accountable state (Nazarbayev 2015). According to the 98th step of this plan, in 2020 the government will introduce a fourth level of budget for towns and villages belonging to the third tier of the subnational government. The town budgets are expected to receive revenue from (i) individual income tax on income not taxable at the source of payment, (ii) tax on the property of individuals, (iii) tax on transport, (iv) land tax, (v) fees for the placement of outdoor (visual) advertising, (vi) fees from renting municipal property, (vii) donations, (viii) fines for administrative violations, (ix) income from the sale of municipal property, and (x) transfers from the district budget. In fact, given the low salaries and poor capital assets, in most cases transfers from district budgets will continue to make up the largest share of the monotowns' budgets. Nowadays, the budgets of monotowns belonging to the second tier of the subnational government are 90% dependent on the regional transfers, except for the oil- and gas-based budget of Zhanatas, where 100% of the budget revenue comes from local taxes.

Although policies permitted monotown authorities to participate in drafting local plans, the latter were not able to develop feasible projects or implement all planned activities due to administrative and fiscal constraints. None of the programmes included measures to facilitate and improve decision-making or reduce the monotowns' current financial dependence. Being at the lowest level of public administration, the town authorities have been unable to deal with serious economic and environmental problems. Monotown authorities were granted subsidies to build new urban infrastructure, but they did not receive adequate budget allocations for the maintenance of such infrastructure. Lacking regulatory and taxation power and adequate managerial capacity, monotown authorities could not engage businesses or motivate large enterprises to participate in the restructuring of the local economy.

Lessons Learned and Ways to Proceed

Planning Aligned with Local Development Needs

Kazakhstan has made several attempts to transform industrial towns by channelling public investment into monotowns. However, government investment has not resulted in considerable economic diversification because the government's top-down approach ineffectively framed monotowns within the concept of dependency on a single large enterprise. The government attempted to introduce a bottom-up approach by allowing town authorities to identify and propose investment projects and design local plans. However, lacking expertise and being constrained by a single methodology, town authorities were mostly unable to formulate realistic development plans or engage private sector actors in an effective dialogue. The delegation of planning was a positive change, but the national government instructions and indicators needed to be better customised to the needs of each monotown, as each urban area had different assets and liabilities. Considering the industrial past and present, there is a need to increase the number of indicators measuring environmental conditions and introduce policies aimed at improving the ecological situation in monotowns. Planning that is closely aligned with specific monotown development needs can significantly improve the implementation of actions designed at the national level to support monotowns.

Strengthening the Capabilities of Town Authorities to Mobilise Locational Advantages

The national government should invest in strengthening the capability of monotown authorities to make evidence-based decisions about local investments to improve their effectiveness. Town authorities need to learn how to focus on feasible small-scale, market-driven investment projects and attract public and private project funding. If the national government wants to facilitate the development of non-mining businesses, it is important to strengthen the abilities of monotown authorities to identify and exploit their towns' locational advantages. International practice suggests that making effective use of locational advantages depends on the town authorities' ability to arrange urban development around the uniqueness of their local resources and the needs of the local population. Local plans cannot be a useful tool when the town authorities are the sole actor responsible for planning and implementation. The success of a local plan depends on how well it represents the long-term vision for the community. The national government should ensure that adequate mechanisms are created enabling town development plans to serve as a tool to involve key public and private actors in the planning and implementation processes.

Better Access to Local Resources and Funding for Responsible Urban Governance

Town development plans will not make much difference if monotown authorities do not have decision-making powers, own budget, or access to other local resources required to successfully implement the plans. If the national government wants to make monotown authorities responsible for town development, it is important to give them adequate power to execute these responsibilities. Monotown authorities should be incentivised to improve residents' living conditions by making use of local assets and deriving benefits from national transfers. The monotown policy enabled towns to use public investment to rehabilitate urban infrastructure, but the infrastructure can deteriorate even in the short run if there is no funding to manage and maintain it. In some monotowns, municipal infrastructure and industrial sites still belong to large enterprises or their respective ministries, and town authorities do not have authority to manage these local resources. In addition, town authorities should be given adequate incentives to create favourable economic conditions and opportunities to generate revenue for reinvestment in town development. However, fiscal decentralisation should be strategic and well planned, considering that most monotowns do not posses the economic resources to generate sufficient revenue to cover all development-related expenses.

Innovative Arrangements for Retaining Young Talent in Monotowns

The government should replace the resettlement to monotowns of repatriates and young people (particularly those with university degrees) with more innovative arrangements, aimed at finding new combinations of national policy efforts to attract young talent particularly to small monotowns. For example, the national government could combine urbanisation and digitalisation policy initiatives and assist international and national companies with job creation in the digital business for monotown dwellers, so that they can work via the Internet without relocating. Youth participation in digital employment could be a new criterion for channelling public investment to monotowns. Inclusion of this criterion might serve as an incentive to attract the public investment required to create online-based jobs. Furthermore, the development of digital entrepreneurial skills could help monotown businesses to enter the national and international markets and increase their contributions to the town economy. Digital tools in the hands of young entrepreneurs may assist monotowns in becoming more visible to wider groups of the population and the global community, so that monotowns could benefit from their locational advantages and tourism potential.

A National Association of Monotowns

In many cases, monotown authorities are best suited to identify and prioritise town development needs. Therefore, the national government should learn to listen to municipalities and understand their development priorities. It may be useful to create horizontal platforms for inter-governmental dialogue and to help monotown authorities and regional and national government to integrate their efforts to tackle local development challenges. For example, monotown authorities could form a national association of monotowns, similar to the Federation of Canadian Municipalities. An association could stimulate joint learning- and experience-sharing among monotown authorities and open up opportunities for cooperation with similar towns worldwide. The national government can support the association of monotowns by creating a special fund for local development. The availability of the fund would allow the distribution of grants on a competitive basis by the decision of all association members. This type of funding can serve as a good alternative to the current distribution of subsidies and increase the efficiency of national support schemes to make monotown authorities more innovative, creative, and accountable to their communities.

Conclusion

The first monotown policy was a government reaction to societal pressure to increase wages and improve living conditions in monotowns as opposed to the *Kazakhstan-2050* strategy that is implemented in a planned fashion. The 2050 strategy calls for the use of the powers of urbanisation and regional development to achieve fast economic growth. In 2012, the government included monotowns in its pro-urbanisation policy by approving regional development programmes. However, the government has maintained its outdated approach and continues to assist monotowns via the top-down distribution of subsidies that are often channelled primarily to uncompetitive and unprofitable industries, ultimately leading to the inefficient use of public investment. This path dependency contradicts the first objective of the 2050 strategy: to ensure economic development based on the principles of profitability, return on investment, and competitiveness. The current monotown policy is not effective for national economic or social development because government interventions create even greater dependency of the monotowns on their local large enterprises.

Most Kazakhstani monotowns play an important role in the economic development of the country by serving the mining industry. However, dependence on mining is an obstacle to sustainable development because natural resources are exhaustible and their extraction leads to unavoidable environmental pollution. Therefore, monotowns must find more sustainable ways of development. If the national government continues to solve monotown development challenges by proposing a standard recipe of top-down subsidies of industry-related development, public money may still be wasted and economic restructuring is unlikely to be achieved. If town authorities remain incapable of implementing ambitious national objectives, the government faces the risk of losing the public trust. In this case, the living conditions in monotowns are likely to remain substantially unchanged. These towns will continue losing human capital including young talent and active firms that will keep moving to larger urban areas that offer better living and business conditions.

Today the national government faces the challenge of designing and implementing national policies that should better meet monotowns' development needs. For actors at all levels of the government, it would be helpful to pay proper attention to the fifth objective of the *Kazakhstan-2050* strategy, which emphasises the need for better governance and democracy at all levels of public administration. The *Kazakhstan-2050* strategy calls for the rational use of natural resources, as well as for strengthening local government. The policy interventions need to better aligned with these two important strategic objectives. Ideally, delegated responsibility for planning must be accompanied by adequate allocation of decision-making power and access to resources to the local authorities. There is a need to consider some degree of decentralisation of power and resource management to enable monotown authorities to cooperate with the local community, support local businesses, and turn national subsidies into the effective public investment for the long-term sustainable development of their town.

According to the *Kazakhstan-2050* strategy, the country needs strong subnational governments. However, the decentralisation path requires certain institutional preconditions. Effective decentralisation can take place if monotown authorities are made accountable to their residents and are incentivised to engage in sustainable economic development. Decentralisation of planning and public investment would be more effective if it transformed the role of monotown authorities from mere implementers of top-down orders to facilitators of bottom-up innovative initiatives. The decentralisation of control over local resources requires monotown residents to become key actors in town development. The decentralisation of revenue and taxation that gives local authorities power over local budgets and the distribution of investments would increase monotown authorities' motivation to cooperate with local industries and small businesses. To conclude, achieving sustainable economic development for the monotowns is possible if the monotown policy is better aligned with the decentralisation objectives of the *Kazakhstan-2050* strategy, which requires gradual changes of the roles and capabilities of local development actors—town authorities, local businesses, and community activists.

Appendix

See Table 8.5.

Table 8.5 Study participants

Participant No	*Gender*	*Age group*	*Affiliation*	*Other details*	*Location*	*Interview date*
1	Male	40–44	National government	Member of staff, Ministry of the National Economy	Nur-Sultan	August 2017
2	Male	40–44	National government	Member of staff, Ministry of Innovative Development	Nur-Sultan	August 2019
3	Male	45–50	Parliament	Deputy	Nur-Sultan	August 2019
4	Male	45–50	Town authorities	Manager, town administration	Tekeli	April 2017
5	Male	30–34	Town authorities	Manager, Youth Centre	Tekeli	April 2017
6	Female	45–50	Town authorities	Member of staff, Internal Policy Department	Tekeli	June 2019
7	Male	40–44	Town authorities	Member of staff, Department of Economic and Budget Planning	Tekeli	June 2019
8	Male	30–34	Town authorities	Member of staff, Department of Entrepreneurship Development	Tekeli	June 2019
9	Male	40–44	Town authorities	Member of staff, Department of Economic and Budget Planning	Kentau	April 2017
10	Male	30–34	Town authorities	Member of staff, Department of Entrepreneurship Development	Kentau	April 2017
11	Male	40–44	Mining industry	Senior manager, Electromanganese LLC	Tekeli	June 2019
12	Male	45–50	Mining industry	Senior manager, Electromanganese LLC	Tekeli	June 2019
13	Male	65–70	Mining industry	Member of staff, Mining Buro LLC	Tekeli	June 2019
14	Male	40–44	Mining industry	Senior Manager, EuroChem JSC	Zhanatas	April 2017
15	Female	30–34	Mining industry	Member of staff, EuroChem JSC	Zhanatas	April 2017

(continued)

Table 8.5 (continued)

Participant No	*Gender*	*Age group*	*Affiliation*	*Other details*	*Location*	*Interview date*
16	Male	40–44	Public organisation	Member of staff, Economic Research Institute	Nur-Sultan	August 2017
17	Female	45–50	Public organisation	Member of staff, Academy of Public Administration	Nur-Sultan	August 2017
18	Female	50–54	Public organisation	Senior manager, a school	Tekeli	April 2017
19	Female	30–34	Public organisation	Senior manager, a school	Tekeli	April 2017
20	Female	25–30	Public organisation	Senior manager, a school	Tekeli	April 2017
21	Male	40–44	Public organisation	Senior manager, House of Culture	Tekeli	April 2017
22	Female	45–50	Non-profit organisation	Manager, National Chamber of Entrepreneurs ATAMEKEN	Nur-Sultan	August 2017
23	Male	40–44	Non-profit organisation	Senior manager of a local branch, National Chamber of Entrepreneurs ATAMEKEN	Tekeli	April 2017
24	Male	40–44	Entrepreneur	Senior manager, a local college	Tekeli	April 2017
25	Female	40–44	Entrepreneur	Manager, tourism training school	Tekeli	April 2017
26	Female	40–44	Entrepreneur	Editor, a local newspaper	Tekeli	April 2017
27	Male	25–30	Entrepreneur	Owner, a local shop	Tekeli	June 2019
28	Female	45–50	Entrepreneur	Owner, a local shop	Kentau	June 2019
29	Male	25–30	Entrepreneur	Owner, a local shop	Kentau	June 2019
30	Female	40–44	Entrepreneur	Owner, a local shop	Khromtau	August 2018
31	Female	25–30	Monotown resident	Housewife	Tekeli	June 2019
32	Male	40–44	Monotown resident	Taxi driver	Tekeli	June 2019
33	Male	40–44	Monotown resident	Taxi driver	Zhanatas	April 2017
34	Male	30–34	Monotown resident	Taxi driver	Kentau	June 2019

Participant No	*Gender*	*Age group*	*Affiliation*	*Other details*	*Location*	*Interview date*
35	Male	25–30	Monotown resident	Taxi driver	Karatau	April 2017
36	Female	40–44	Monotown resident	Worker, a local street market	Tekeli	June 2019
37	Female	45–50	Monotown resident	Worker, a local street market	Zhanatas	April 2017
38	Male	40–44	Monotown resident	Worker, a local street market	Kentau	June 2019
39	Female	20–24	Monotown resident	Worker, a local shop	Tekeli	June 2019
40	Female	40–44	Monotown resident	Worker, a local shop	Tekeli	June 2019
41	Female	20–24	Monotown resident	Worker, a local shop	Zhanatas	April 2017
42	Male	30–34	Monotown resident	Worker, a local shop	Temirtau	April 2017
43	Male	20–24	Monotown resident	Worker, a local shop	Khromtau	August 2018
44	Female	25–30	Monotown resident	Worker, a local shop	Karatau	April 2017
45	Female	30–34	Monotown resident	Hairdresser	Tekeli	June 2019
46	Male	25–30	Monotown resident	Hairdresser	Temirtau	April 2017
47	Female	40–44	Monotown resident	Homemaker	Tekeli	June 2019
48	Female	50–54	Monotown resident	Homemaker	Tekeli	June 2019
49	Female	25–30	Monotown resident	Homemaker	Tekeli	June 2019
50	Female	30–34	Monotown resident	Homemaker	Kentau	June 2019
51	Female	25–30	Monotown resident	Homemaker	Temirtau	April 2017
52	Female	40–44	Monotown resident	Homemaker	Khromtau	August 2018

References

Akhmediyeva, A. (2012, September 6). *Ashes of 12 Japanese prisoners sent home from Tekeli* [in Russian]. Radio Azattyk. Available via https://rus.azattyq.org/a/ashes-of-japonese-prisoners-of-war-sent-from-kazakhstan/24698886.html. Accessed 16 January 2020.

Akimat of Arkalyk. (2016). *Programme for the development of territory of Arkalyk, Kostanay region, for 2016–2020* [in Russian]. Available via http://arkalyk.kostanay.gov.kz/ekonomika-i-byudzhet/programma-razvitiya-territorii-goroda-arkalyka-na-2011-2015-gody/. Accessed 27 April 2020.

Akimat of Balkhash. (2016). *Programme for the development of territory of Balkhash for 2016–2020* [in Russian]. Available via http://www.balkhash.gov.kz/ru/o-gorode_1. Accessed 27 April 2020.

Akimat of Rudny. (2016). *Programme for the development of territory of Rudny for 2016–2020* [in Russian]. Available via http://rudeconom.kz/index.php/ru/programma-razvitiya-territorii-goroda-rudnogo. Accessed 27 April 2020.

Akimat of Rudny. (2019). *History of Rudny town* [in Russian]. Available via http://rudny.kostanay.gov.kz/o-gorode/istoriya-obrazovaniya-goroda/. Accessed 16 January 2020.

Akimat of Stepnogorsk. (2017). *Programme for the development of territory of Stepnogorsk for 2016–2020* [in Russian]. Available via http://stepnogorsk.gov.kz/content/programma-razvitiya-territoriy-na-2016-2020-gody. Accessed 27 April 2020.

Akimat of Temirtau. (2020). *About Temirtau* [in Russian]. Available via http://temirtau-akimat.gov.kz/ru/about/. Accessed 16 January 2020.

Akimat of Zhanatas. (2016). *Zhanatas: yesterday, today and tomorrow. Video presentation of the Master Plan of Development of the town of Zhanatas* [in Russian]. Available via https://www.youtube.com/watch?v=VwEqoqcqfPE. Accessed 27 April 2020.

Aldashev, A., & Dietz, B. (2014). Economic and spatial determinants of interregional migration in Kazakhstan. *Economic Systems, 38*(3), 379–396.

Beisembaev, S., Insebaeva, S., & Moldokanov, D. (2017). *'Neither to the village, nor to the city': The problem of informal youth employment in single-industry towns of Kazakhstan* [in Russian]. Astana: Paper Lab.

BISAM Central Asia. (2012). *Monotowns of East Kazakhstan: situation, problems, solutions* [in Russian]. Almaty: BISAM Central Asia. Available via https://www.kz.undp.org/content/kazakhstan/ru/home/library/democratic_governance/_-_-_-_-_-_1111.html. Accessed 28 March 2020.

Central Asia Monitor. (2012, May 3). *Going far, receiving less* [in Russian]. Available via https://camonitor.kz/3819-.html. Accessed 18 February 2020.

Committee on Statistics under the Ministry of the National Economy of the Republic of Kazakhstan (Committee on Statistics of Kazakhstan). (2020).

Regions of Kazakhstan 1991–2018 [in Russian]. Available via http://www.stat.gov.kz/. Accessed 10 January 2020.

Didyk, V. V., & Ryabova, L. A. (2014). Single-industry towns of the Russian Arctic: Development strategy using the case study of the cities/towns in the Murmansk Oblast [in Russian]. *Economic and Social Changes: Facts, Trends, Forecast, 4*(34), 22–34.

Electromanganese. (2020). *History of plant* [in Russian]. Available via http://www.emn.kz/o-predpriyatii. Accessed 22 December 2020.

Government Resolution of the Republic of Kazakhstan No 1187. (1996). *Resolution on additional measures to remove the Tekeli Lead Zinc Plant from the crisis* [in Russian]. Available via http://adilet.zan.kz/rus/docs/P960001187_/links#to. Accessed 19 October 2019.

Government Resolution of the Republic of Kazakhstan No 1096. (2003). *Resolution on Approval of Strategies of industrial and innovative development of the Republic of Kazakhstan for 2003–2015* [in Russian]. Available via https://online.zakon.kz/document/?doc_id=1039961. Accessed 2 October 2019.

Government Resolution of the Republic of Kazakhstan No 958. (2010). *Resolution on the Approval of the Programme for Fast Industrial and Innovative Development of the Republic of Kazakhstan for 2010–2014* [in Russian]. Available via https://tengrinews.kz/zakon/prezident_respubliki_kazahstan/promyishlennost/id-U100000958_/. Accessed 12 March 2020.

Government Resolution of the Republic of Kazakhstan No 683. (2012). *Resolution on the Approval of Monotown Development Programme for 2012–2020* [in Russian]. Available via http://adilet.zan.kz/rus/docs/P1200000683. Accessed 11 November 2019.

Government Resolution of the Republic of Kazakhstan No 728. (2014). *Resolution on Approval of Regional Development Programme 2020* [in Russian]. Available via http://online.zakon.kz/Document/?doc_id=31584094#pos=29;-284. Accessed 3 February 2020.

Government Resolution of the Republic of Kazakhstan No 15. (2015). *On amendments to the Government Resolution No 728 on approval of the Regional Development Programme 2020* [in Russian]. Available via https://online.zakon.kz/m/document/?doc_id=31664529. Accessed 27 April 2020.

Government Resolution of the Republic of Kazakhstan No 42. (2015). *On amendments to the Government Resolution No 728 on approval of the Regional Development Programme 2020* [in Russian]. Available via https://online.zakon.kz/service/doc.aspx?doc_id=38381861. Accessed 27 April 2020.

Government Resolution of the Republic of Kazakhstan No 284. (2015). *On amendments to the Government Resolution No. 728 on approval of the Regional Development Programme 2020* [in Russian]. Available via https://zakon.uchet.kz/rus/docs/P1500000284. Accessed 27 April 2020.

Government Resolution of the Republic of Kazakhstan No 451. (2015). *On amendments to the Government Resolution No. 728 on approval of the Regional Development Programme 2020* [in Russian]. Available via https://online.zakon.kz/document/?doc_id=31584094#pos=0;0. Accessed 27 April 2020.

Government Resolution of the Republic of Kazakhstan No 499. (2016). *On amendments to the Government Resolution No 728 on approval of the Regional Development Programme 2020* [in Russian]. Available via https://online.zakon.kz/document/?doc_id=36912352. Accessed 27 April 2020.

Government Resolution of the Republic of Kazakhstan No 767. (2018). *On approval of the State Programme of Regional Development until 2020* [in Russian]. Available via https://online.zakon.kz/Document/?doc_id=36463121. Accessed 27 April 2020.

Government Resolution of the Republic of Kazakhstan No 990. (2019). *Resolution on Approval of State Regional Development Programme 2020–2025* [in Russian]. Available via https://www.zakon.kz/5003230-utverzhdena-gosudarstvennaya-programma.html. Accessed 19 February 2020.

Kim, P. G. (1993). *Koreans of Republic of Uzbekistan* [in Russian]. Tashkent: Uzbekistan.

Kurmanov, A. (2016). *Kazakhstan tragedy. Five years since the shelling of Zhanaozen workers*. Available via https://www.fergananews.com/articles/9195. Accessed 14 February 2020.

Ministry of the Economic Development and Trade. (2012). *Monotown Development Programme*. Presentation [in Russian]. Available via https://www.slideshare.net/Atameken/ss-12220137. Accessed 27 April 2020.

Nazarbayev, N. (2012). *Strategy Kazakhstan-2050: New political course of the established state*. Available via https://strategy2050.kz/en/page/multilanguage/. Accessed 17 January 2020.

Nazarbayev, N. (2015). *National Plan—100 concrete steps to implement the five institutional reforms of the Head of State Nursultan Nazarbayev* [in Russian]. Available via https://online.zakon.kz/document/?doc_id=31977084. Accessed 17 January 2020.

Organisation for Economic Co-operation and Development (OECD). (2017). *Urban policy review: Kazakhstan*. Paris: OECD Publishing. Available via https://www.oecd.org/publications/oecd-urban-policy-reviews-kazakhstan-9789264274624-ru.htm. Accessed 15 February 2020.

Statistics Department of the Eurasian Economic Committee. (2020). *National accounts data*. Available via http://www.eurasiancommission.org/ru/act/integr_i_makroec/dep_stat/econstat/Pages/national.aspx. Accessed 15 February 2020.

Suilemanova, A., Seilkhanov, T., & Akhatila, R. (2017). *Zhanatas will become a city with a successfully developing industry* [in Russian]. Available via https://24.kz/ru/news/social/item/160965-zhanatas-stanet-gorodoms-uspeshno-razvivayushchejsya-promyshlennostyu. Accessed 29 September 2019.

Trotchenko, P. (2018). *Kentau: How does a single-industry town live?* [in Russian]. Radio Azattyk. Available via https://rus.azattyq.org/a/kentau-chem-jivet-monogorod/29467563.html. Accessed 16 January 2020.

Vechor-Shcherbovich, S. (1975). *Planning of manpower in the Soviet Union* [in Russian.]. Moscow: Progress Publishing.

Zholdybayev, Zh., & Zhamelov, B. (2007). *The role of small towns in the socio-economic development of South Kazakhstan* [in Russian]. Available via http://www.rusnauka.com/8_NMIV_2013/Istoria/2_131351.doc.htm. Accessed 3 November 2019.

PART III

Conclusion

CHAPTER 9

Insights into Embedded Policy Paradigms and Kazakhstan's Future Trajectory

Nikolai Mouraviev

Introduction

This book provides an assessment of policies in different sectors of Kazakhstan's economy, how policies have changed over time, since 1991, and how successfully they transformed the sectors. The collection also offered insights into policy areas beyond those in specific sectors. These areas include procedural justice in public service delivery, water–energy–food (WEF) nexus governance, public–private partnerships (PPPs) as a tool for public management, and policy development for monotowns. How do conclusions drawn from the book's chapters link to each other? What can one learn from them? Importantly, what are the implications for Kazakhstan's future policy and governance in the light of the ambitious *Kazakhstan-2050* strategy that is on the nation's agenda since its adoption in 2012? This chapter aims to answer these questions, at least in part.

N. Mouraviev (✉)
Abertay University, Dundee, UK
e-mail: n.mouraviev@abertay.ac.uk

A. Koulouri and N. Mouraviev (eds.),
Kazakhstan's Developmental Journey,
https://doi.org/10.1007/978-981-15-6899-2_9

While investigating policies and their implementation in various parts of Kazakhstan's economy, as well as intersectoral practice, a number of ideas and themes have been reiterated by the authors throughout the book. The overarching theme is sustainability; that is, policy and governance have to ensure the nation's sustainable development. In the book, this theme has been discussed in detail, although at times implicitly, making use of a large variety of analytical tools and concepts, with regard to many sectors and policy areas, and Kazakhstan's economy in general. This is hardly surprising: every country needs to ensure sustainability, although in every country a set of resource constraints, historical legacy, and broader political and economic reality shape its developmental pathway. In Kazakhstan, although since 1991 the nation achieved remarkable progress in many fields by taking advantage of the sale of its vast natural resources, the future of some industries does not appear to be resting on policies that are best suited to serving the *Kazakhstan-2050* strategy, and would ensure sustainability in the twenty-first century.

In addition to sustainability, the book chapters emphasise other themes that can be viewed as the means to achieve sustainability. They may be summarised as calls to ensure:

- economic restructuring;
- participatory governance;
- intersectoral thinking, rather than departmentalisation; and
- policy integration, rather than fragmentation.

It is worth noting that the themes are overlapping, and the means to achieve sustainability are to a certain degree interdependent. For example, participatory governance might not only better inform policy and, as a consequence, ensure better outcomes, but also contribute to policy integration, and intersectoral policymaking.

What do the themes' narratives tell us about the suitability of the nation's policies to meet the challenges of the twenty-first century? Is the country on track to achieve the objectives set in the *Kazakhstan-2050* strategy, or is it necessary to shift some policies? Do the ideas within the themes (that could be viewed as recommendations for improvement) align with the *Kazakhstan-2050* strategy? If there is misalignment, there might be a mismatch between the achieved result and the strategic objectives set. The ideas put forward in the book go beyond the boundaries of a specific industry. They should not be viewed as a plan of action with defined resource allocation, set timeline, and clearly delineated expected outcomes. Rather, the ideas highlight elements of the policies that could be shaping Kazakhstan's economy and society in general for years ahead. The *Kazakhstan-2050* strategy leaves significant room for identifying ways to achieve its objectives, and this book takes advantage to offer insights into what these ways might be. At the policy level, most of the book's ideas are conceptual, although occasionally (e.g. regarding agriculture, education, or PPPs) the book also presents a practitioners' perspective. Each theme will now be discussed in greater detail.

Sustainability as an Overarching Theme

In many chapters, the sustainable development concept transpires indirectly, as a result of analysing specific conditions, or changes in a sector. For example, the study of agricultural reforms over 30 years, since 1991, showed that only in the twenty-first century environmental and social dimensions became part of the policy objectives in this sector, and it took about ten years (beginning approximately in 2010) for various actors to start paying adequate attention to these dimensions (see Chapter 3). One might argue that in the 1990s, during the most turbulent time of transition to a market system, policymakers naturally focused on the sector's survival; that is, its economic durability. Although this might be true, the focus on survival lacked the accompanying development of market relations in agriculture. In the light of expanding market relations, the use of instruments suitable for a market economy would have been more appropriate.

Despite significant improvements over three decades, agriculture has been and still is relying on government subsidies, which is hardly a tool that supports market relations. Rather, a subsidy, whether in agriculture or any other sector, typically indicates and/or even reinforces a producer's inefficiency, as the subsidy pays for part of the production costs. As this book emphasises, the economic sustainability of agriculture depends on developing market relations, which would allow to attract private investment and, over time, reduce and ultimately phase out subsidies.

However, it is worth recognising the trade-off between a policy aiming at developing efficient, competitive, and overall sustainable agriculture, on the one hand, and social policy supporting the rural population and their livelihood, on the other. As subsidies ensure continuity of agricultural producers' livelihood and employment in rural areas, one would agree that terminating subsidies across the sector in the near future is neither realistic, nor desirable. However, in the long run, agriculture cannot rely on subsidies as a policy paradigm and, in the future, efficiency considerations may prevail. The solution to this dilemma (efficiency versus social support) may lie in a paradigm shift, when at least two approaches may be implemented at the same time: policymakers might consider a gradual, incremental phasing-out of subsidies over a long period (10–20 years) for most cost-inefficient products, and also implement novel, innovative methods in agriculture, such as cluster development based on deploying renewable energy-generating facilities.

Another area in which sustainability is often discussed is the energy sector. The debate about the role of fossil fuels in Kazakhstan's sustainable development often transforms into a discussion on energy security. Over 30 years, policymakers emphasised security of supply as the energy sector's primary objective. Drawing on their own research in Kazakhstan, the editors of this book offered, in earlier publications, a novel conceptualisation of energy security, which is particularly applicable to resource-rich nations and aligns well with the concept of sustainability, and the objectives set in the *Kazakhstan-2050* strategy (Koulouri and Mouraviev 2019; Mouraviev and Koulouri 2019a, b). Energy security could be viewed as a sustainable use of renewables, which means the purposeful and ever-increasing utilisation of renewable sources for the production of power, and a corresponding decrease in reliance on non-renewable sources, complemented by resource efficiency programmes (Koulouri and Mouraviev 2019). Owing to climatic conditions, Kazakhstan has significant potential for renewable energy generation, and over the past ten years the government has designed legal and institutional frameworks for the promotion of renewables. However, high dependence on fossil fuels persists, while the share of renewables in the nation's energy mix remains very small (see Chapter 7 about embedding WEF nexus in policymaking). The country might significantly benefit from adopting a new energy security concept, which would ensure the paradigm shift required for long-term sustainable development based on continuous increases in the utilisation of renewables (Mouraviev and Koulouri 2018).

Why does sustainability often appear a subtle, rather than a pronounced, policy perspective? This occurs because some policies lack coherence or might be contradictory, which does not allow sustainability objectives to be achieved (in addition to Chapter 3 on agricultural policy, see, for example, Chapter 2 on the oil sector, Chapter 6 on PPPs, and Chapter 8 on monotowns). Therefore, sustainability as a concept unifying economic, social, and environmental dimensions of development in a field is yet to be embedded in policymaking and governance.

Theme 1: Ensure Economic Restructuring

Economic restructuring is relevant to the energy sector (Chapter 2), agriculture (Chapter 3), monotowns policy (Chapter 8), and, in fact, the entire economy. In addition to investigating conditions in the oil sector, this book draws on studies in agriculture, energy, nexus governance, and monotowns to emphasise the importance of the economic restructuring that Kazakhstan has been trying to accomplish for many years, although with limited success.

It is worth emphasising the use of the term restructuring as opposed to diversification. At the corporate level, diversification typically means adding a new product to the company's portfolio, which usually pursues the goal of expansion to a new market and/or of spreading the risk over a broader range of products, so that any particular product's exposure to risk is reduced (Baysinger and Hoskisson 1989; Miller 2004; Guerras-Martína et al. 2020). Similarly, at the regional or national level, diversification would usually mean a larger variety of assets and/or industries. However, does a larger variety of industries ensure reduced risk of dependency on fossil fuels? Not necessarily, as each of the newly added industries might bring a tiny, or no tax revenue to the budget. After 30 years of independent economic development, the country has a range of industries, such as the manufacturing of furniture, clothing, electrical equipment, as well as services provided by small businesses, but they do not bring any significant budget revenue. Therefore, diversification as an objective in policy documents or government programmes becomes a somewhat misleading goal, as it might be emphasising the need to create a new industry without necessarily reducing the size of the oil sector and, hence, the corresponding budget dependency on oil revenue.

A new paradigm appropriate for the *Kazakhstan-2050* strategy requires restructuring of the economy, which means investment in new sectors, and their development to the extent that they bring sizeable tax revenue. Ultimately, this would mean a corresponding decrease in dependence of the budget on oil revenue. In this manner, the economy would be increasingly relying on the new sectors, while gradually reducing its dependency on the oil industry. As an additional illustration of what needs to be done, it is worth emphasising that restructuring is not about just creating new sectors or sub-sectors, for example in agriculture. The objective is to ensure that both new and existing sectors and sub-sectors are profitable, and therefore do not require subsidies and, ideally, yield tax revenue to the budget.

The restructuring theme is also critical for the discussion of the oil sector and its role in Kazakhstan's economy. Although the government skilfully implemented its energy policy from the 1990s to the present to ensure that vast natural reserves served well to secure budget revenue for the country's development, there are growing concerns about the depletion of fossil fuels, and the ecological damage resulting from the use of oil, gas, and coal (see Chapter 2). More broadly, concerns are mounting regarding whether the oil sector's and the nation's overall developmental pattern are economically and environmentally sustainable in the long run. Will Kazakhstan be able to join the ranks of the 30 most developed nations, as stipulated by the *Kazakhstan-2050* strategy, amid uncertainty about economic and ecological sustainability, which inevitably undermines the social dimension of the country's development? The calls for a paradigm shift, ensuring gradual departure from the country's reliance on fossil fuels as the dominant source of government revenue, to a restructured economy that would provide stable and significant tax revenue streams from many sources, rather than one, are not new. However, over time these calls intensified and will be intensifying further as long as restructuring remains slow or stalled.

On a smaller scale, the concept of restructuring is pertinent to the development of monotowns. Although this policy area is where for many years the focus was on diversification, minimal results have been received in terms of reducing the towns' dependency on a single large enterprise (see Chapter 8). Some diversification was achieved, however it created insignificant additional employment opportunities, while the historically formed dominance of a single major local employer remains in many towns. As Chapter 8 shows, efforts to develop small businesses failed to change the employment structure in small towns. The deeply embedded policy paradigm that does not acknowledge the towns' economic restructuring as a primary objective requires a shift, which would need to align with broader policy objectives for restructuring the entire economy.

Theme 2: Ensure Participatory Governance

A few chapters raised concerns about path dependency in policymaking when the traditional top-down approach dominates, while citizens and/or local governments are not invited to participate. In the political reality of Kazakhstan with centralised decision-making, it would be unwise to call for completely different methods of policy design. Nonetheless, strengthening the nation's statehood and achieving a certain degree of decentralisation are viewed by the *Kazakhstan-2050* strategy as critical objectives that would facilitate the country's competitiveness. Without replacing the centralised approach to decision-making, this could be achieved by participatory governance, resting on much greater engagement of citizens, local authorities and organisations representing relevant industries. The input of the local-level actors would offset the deficiencies of top-down policy design and implementation, and would complement the centralised model (Chhotray and Stoker 2009; Gustafson and Hertting 2017).

This need for participatory governance is vividly observed in such areas as higher education (Chapter 4), and monotown development (Chapter 8). For example, in higher education many of Kazakhstan's universities are prepared to provide significant input in how the learning processes and research need to be managed. To this end, they would benefit (as the international best practice shows) from more autonomy and greater accountability, which might transform the educational sector based on the application of participatory governance, and, importantly, ensure education quality at the level of the most advanced economies in the world.

In another example, path dependency in policymaking using the traditional top-down approach set serious limitations to the development of monotowns. As Chapter 8 explains, many monotowns experienced continuous decline over the past 30 years, show high levels of unemployment, and persistent outflow of the labour force. These towns will not be able to transform their economy without significant investment by the national government. However, local authorities, rather than the national-level governing bodies, are best equipped to assess their territories' needs, design development plans, and implement investment programmes. A participatory approach would not only better satisfy the local needs, but might also lead to the much-desired economic restructuring on a broader (regional or national) scale (Cornwall 2008). A possible paradigm shift in the direction of participatory governance could therefore be viewed as a realistic possibility to transform policymaking in education, monotown development, as well as agriculture and other sectors of the economy.

In another area, WEF nexus, the need for participatory governance is particularly vivid as the nexus concept is fairly novel for Kazakhstan (see Chapter 7). Participatory governance strongly links to stakeholders' relational equity (for details, see Koulouri and Mouraviev 2020), which emphasises the critical importance of engaging actors, seeking and managing their contributions to the overlapping sectors in order to apply an integrated approach to policy and governance. In turn, this might deliver better outcomes regarding economic sustainability of all three sectors, and ensure environmental and social sustainability (Mouraviev and Koulouri 2020).

Theme 3: Ensure Intersectoral Thinking, Rather Than Departmentalisation

Departmentalisation might be viewed as a Soviet legacy and a feature of time, from the 1990s to present, when the country focused on transforming its key sectors from the command system to a market-based one, a transformation currently manifested as an embedded paradigm in many policy areas. Building a competitive economy with the goal to become one of the 30 most developed nations requires a different approach, in which intersectoral thinking would provide coherent, rather than isolated, solutions, with spillover benefits for the entire economy, rather than just serving sectoral priorities.

The theme of intersectorality has been discussed in all chapters, and illustrated by a range of examples of narrowly focused policy actions and governance solutions that constrained the overall effectiveness of programmes, impeded the achievement in full of the stated objectives, and led to unforeseen negative impacts in other policy areas. Sector-focused decisions have been observed for many years in the oil industry, agriculture, water management, small town and rural development, higher education, transport, and utilities infrastructure, often without any attempt to identify the consequences of isolated decisions on other sectors. For example, the current minimal application of the WEF nexus concept to the governance of the agriculture, water, and energy sectors serves as a most vivid illustration of isolated sector-focused policymaking (see Chapter 7). Another example is in the field of PPP governance, where poorly integrated decisions have led, in recent years, to a disproportionally large number of investment projects, labelled as a PPP, being deployed, significantly increasing the potential government liability if some risks (e.g. insufficient revenue streams, or construction failures) materialise (see Chapter 6).

Theme 4: Ensure Policy Integration, Rather Than Fragmentation

Fragmentation might considerably reduce the effectiveness of governance, and it is not surprising that a number of chapters in this book emphasised the need for effective governance. However, effectiveness is not just about "getting things done". Policy and its implementation need to be holistic, where different parts support each other and ensure consistency. This is not always the case in Kazakhstan. For example, Chapter 2 notes that the world demand for oil will be increasing in the coming years (of course, after the COVID-19 virus is defeated) as many nations will require more energy resources to serve their needs in a time of economic growth, and Kazakhstan intends to take advantage of this by selling more oil. Is this intent part of the embedded policy paradigm that focuses on continuous and ever-increasing sales of fossil fuels? If so, how will this be reconciled with the strategic intent to increase energy generation from renewable sources and reducing the country's dependency on oil revenue?

Another example highlighting the need for policy integration comes from agriculture. In recent years, one of the new policy objectives adopted for this sector stipulated the development of export orientation of selected producers and products, such as wheat. As a result, the country exports wheat, cabbage, carrots, potatoes, buckwheat, sugar, and a range of other agricultural products. Although this might serve as an incentive to improve product quality and, in the long term, increase competitiveness, the export orientation is supported by significant budget expenditure for agriculture. Does export oriented policy have the capacity to continue if budget spending is reduced? The agricultural policy, therefore, might be streamlined and clearer when its different parts are better aligned with each other.

In higher education, although part of the policy supports greater autonomy of universities, which means greater faculty involvement in how academic programmes are formed and run, the other part requires universities to follow the nation-wide centralised regulations that take away flexibility (see Chapter 4). For example, in each taught course (i.e. a semester-long module) there should be two units of assessment—a mid-term exam and a final exam—regardless of the module design, learning objectives, and teaching methods.

The holistic approach to governance has yet another dimension. Investigating the process of public service provision, Chapter 5 argued for the need to ensure procedural justice, rather than just using numbers to show how many services have been provided, and how many citizens have been served. Procedural justice, therefore, refers to the need for governance to be responsive, equitable, and inclusive, although currently the public service delivery seems to focus primarily on efficiency. Treating governance holistically, rather than paying greater attention to one aspect and downplaying the other, is an important point, which is yet to be fully embedded in Kazakhstan's practice.

References

Baysinger, B., & Hoskisson, R. E. (1989). Diversification strategy and R&D intensity in multiproduct firms. *Academy of Management Journal, 32*(2), 310–332.

Chhotray, V., & Stoker, G. (2009). *Governance theory and practice: A cross-disciplinary approach*. London: Palgrave Macmillan.

Cornwall, A. (2008). Unpacking "participation": Models, meanings and practices. *Community Development Journal, 43,* 269–283.

Guerras-Martína, L. Á., Ronda-Pupob, G. A., Zúñiga-Vicentea, J. Á., & Benito-Osorioa, D. (2020). Half a century of research on corporate diversification: A new comprehensive framework. *Journal of Business Research, 114,* 124–141.

Gustafson, P., & Hertting, N. (2017). Understanding participatory governance: An analysis of participants' motives for participation. *The American Review of Public Administration, 47*(5), 538–549.

Koulouri, A., & Mouraviev, N. (2019). Energy security through the lens of renewable energy sources and resource efficiency. In N. Mouraviev & A. Koulouri (Eds.), *Energy security: Policy challenges and solutions for resource efficiency* (pp. 9–35). London: Palgrave Macmillan.

Koulouri, A., & Mouraviev, N. (Eds.). (2020). *Policy and governance in the water-energy-food nexus: A relational equity approach*. Abingdon and New York: Routledge.

Miller, D. J. (2004). Firm's technological resources and the performance effects of diversification: A longitudinal study. *Strategic Management Journal, 25*(11), 1097–1119.

Mouraviev, N., & Koulouri, A. (2018). Clean energy and governance challenges. *International Journal of Technology Intelligence and Planning, 12*(1), 1–5.

Mouraviev, N., & Koulouri, A. (Eds.). (2019a). *Energy security: Policy challenges and solutions for resource efficiency*. London: Palgrave Macmillan.

Mouraviev, N., & Koulouri, A. (2019b). Renewable energy and resource efficiency: Governance is key. In N. Mouraviev & A. Koulouri (Eds.), *Energy security: Policy challenges and solutions for resource efficiency* (pp. 255–261). London: Palgrave Macmillan.

Mouraviev, N., & Koulouri, A. (2020). The water-energy-food nexus: Lessons for governance. In A. Koulouri & N. Mouraviev (Eds.), *Policy and governance in the water-energy-food nexus: A relational equity approach* (pp. 161–165). Abingdon and New York: Routledge.

Index

A. Koulouri and N. Mouraviev (eds.), *Kazakhstan's Developmental Journey*, https://doi.org/10.1007/978-981-15-6899-2

F

The manufacturer's authorised representative in the EU is Springer Nature Customer Service Centre GmbH, Europaplatz 3, 69115 Heidelberg, Germany. If you have any concerns regarding our products, please contact ProductSafety@springernature.com

Printed and bound by CPI Group (UK) Ltd, Croydon, CR0 4YY
30/09/2026
02225502-0009